Radiation-Chemical Processes in Solid Phase

Theory and Application

Evginiy I. Grigoriev
and
Leonid I. Trakhtenberg
Karpov Institute of Physical Chemistry
Moscow, Russia

CRC Press
Boca Raton New York London Tokyo

Radiation-Chemical Processes in Solid Phase

Theory and Application

Aquiring Editor: Svetlana Landau
Contact Editor: Mark Licker
Project Editor: Jennifer Richardson
Marketing Manager: Greg Daurelle
Direct Marketing Manager: Arline Massey
Cover Designer: Denise Craig
Pre-Press: Kevin Luong, Carlos Esser

Library of Congress Cataloging-in-Publication Data

Grigoriev, E. I.
Radiation-chemical processes in solid phase : theory and application / E. I. Grigoriev, L. I. Trakhtenberg.
p. cm.
Includes bibliographical references and index.
ISBN 0-8493-9436-8 (alk. paper)
1. Radiation chemistry. 2. Solids--Effect of radiation on.
I. Trakhtenberg, L. I. (Leonid Izrailevich) II. Title.
QD642.G75 1996
541.3′82—dc20

96-13918
CIP

International Standard Book Number 0-8493-9436-8
Library of Congress Card Number 96-13918
Printed in the United States of America 1 2 3 4 5 6 7 8 9 0
Printed on acid-free paper

Preface

Exactly a century ago the chemical action of radiation was discovered. In 1896, the French physicist Becquerel detected that a photographic plate had been exposed to high energy radiation of uranium salts. Thus, the first radiation-chemical process and spontaneous radioactivity were discovered simultaneously.

Since then much attention has been paid to the chemical aspects of impact of the radiation on a substance. This is due to several reasons. First, some practically important technological processes can be conducted, comparatively inexpensively, under the action of ionizing radiation. Second, the use of different materials under radiation conditions (for example, in nuclear power engineering and in space engineering) leads to the necessity of solving the problem of protecting equipment and living organisms. The efficiency of all these processes is determined by different stages of radiolysis, many of them proceeding in the solid phase.

Thus, it follows that the study of elementary processes of solid-phase radiation chemistry constitutes a very important scientific problem. The radiation chemistry of condensed matter is not studied as well as that of gases. This is easy to understand since in the first case the regularities of radiation-chemical processes are influenced to a great extent by the medium, which is not the case for gases, especially rarefied ones.

It is clear that radiation-induced, solid-state, physico-chemical processes should be studied in terms of solid state physics. It is this approach that has been suggested;[1] however, the Russian series *Physical Chemistry. Modern Problems* had a short run and was not translated into English. That is why the approach proposed by the authors in this publication was not accessible to a wider audience.

Further studies of regularities governing solid-state, radiation-chemical transformations have shown the efficiency of taking into account the dynamic, structural, and relaxing properties of a medium. In some cases, this leads only to small corrections of the mechanism of a chemical process. But, sometimes the properties of a solid play a decisive role, and this should be taken into account while analyzing experimental data. Such processes include energy, charge, and neutral particle transfer, which underlie the radiation chemistry of solids.

Experimental investigations are rather difficult to conduct due to the fact that a radiation-chemical process consists of a sequence of physical and chemical stages (such as primary processes of radiation-matter interaction, relaxation and transfer processes, and mono- and bimolecular reactions), each of which influences the process as a whole. The rates of different stages differ by several orders of magnitude and, consequently, they cannot be studied in one experiment. Data reported so far are not always applicable, since studies are conducted in incomparable conditions. We should note, however, that results of photochemical research have proven useful.

A large number of short-lived chemically active particles and excitations are generated in a substance under irradiation. Their detection is not easy but is necessary

[1] Pshezhetskii, S.Ya., Trakhtenberg, L.I., Radiation-chemical processes in solid phase, in *Physical Chemistry. Modern Problems*, Klotirkan, Ya. M., Ed., Khimiya, Moscow, 1984.

for constructing the general picture of a radiation-chemical process. One possible way to study radiation-chemical transformations in detail is by low and super-low temperatures. This makes it possible to stabilize intermediate products of radiolysis, which are unstable under normal conditions. Also, the intermediate products can be subjected to different factors (light, for example), thus making the radiolysis process flow via another channel. Temperature intervals for which the intermediate products enter the chemical process differ for different particles and so radiation-chemical reactions can be made to proceed selectively. The data obtained can be used to elucidate the radiolysis mechanism at normal temperatures.

Experimental and theoretical findings are now at such a qualitative and quantitative level that there is a need for a monograph in which the media impact is traced from primary stages to the formation of end products.

Chapter 1 starts with a brief description of the notions of the solid-state theory and molecular physics, which facilitates one's perception of the material. Chapter 2 is typical of all monographs on radiation chemistry; it is devoted to radiation-substance interaction. Emphasis here is on chemically active elementary excitations and on energy absorption by homogeneous and heterogeneous systems. Primary radiation-chemical processes are considered in Chapter 3. The next three chapters discuss mechanisms and kinetics of various solid-phase chemical processes. Chapter 7 discusses the joint action of ionizing radiation and light on a substance. Some heterogeneous radiation-chemical processes are dealt with in Chapter 8. Finally, Chapter 9 presents application aspects of solid-phase radiation chemistry. Here, along with traditional applications, such novel problems as radiation impact on high-temperature superconductivity and complex-formation ability of new compounds — C_{60} are considered. A new original method developed by the authors of utilizing radioactive waste is described.

Contents

Acknowledgments

The authors are greatly indebted to Profs. V.I. Goldanskii, S.Ya. Pshezetskii, and D. Stehlik for helpful discussions and suggestions. The authors also wish to thank Drs. V.A. Chernoplekova, A.E. Grigoriev, and P.S. Vorontsov for their help in preparation of the manuscript. Special thanks are extended to Prof. G.N. Gerasimov and Mrs. S. L. Trakhtenberg for their help in writing Sections 4.3 and 8.2., and to E.A. Nasledukhov for his help in the translation of the manuscript. Also, thanks go to Svetlana Landau, CRC acquiring editor, for her help with the production of this book.

1 GENERAL IDEAS OF THE SOLID-STATE THEORY AND MOLECULE PHYSICS

The problems discussed in this chapter are related to processes occurring in an irradiated solid substance. Naturally, properties of the solid phase affect various characteristics of radiation-chemical reactions. In particular, the properties of a medium often determine a reaction mechanism. Therefore, it is necessary first of all to describe a "territory" which is exposed to radiation and in which a chemical process takes place.

Below we shall dwell on the following traditional issues of solid-state physics: band structure and phonon spectrum of the crystalline lattice and energy transfer process (intra- and intermolecular) in volume and surface. In addition, various mechanisms of excited molecules dissociation are considered.

The authors did not set a goal of providing a comprehensive description of the phenomena treated in this chapter. All these issues have been explained in full in published monographs.[1-17] Here only a listing of basic notions is given which a researcher working in the field of solid-phase radiation chemistry is necessarily faced with.

1.1 THE BAND STRUCTURE

A solid (crystalline or amorphous) consists, as is known, of a large number of particles. Depending on the approximation used and the problem of interest to us, it can be molecules, atoms, ions, nuclei, and electrons. But, in any case, the number of variables which correspond to them is macroscopically great. As a result, the Schrodinger equation, by which the electronic structure of a solid body is described, does not have a precise analytical solution. For this reason, a quantum-mechanical description of electronic properties of a solid must be based on a sequence of simplifications. The band theory seems to be the simplest scheme accounting for the most important specific features of electron behavior in crystals.

The theory is based on the following main concepts:

1. Atomic nuclei, due to their large mass, are considered as stationary sources of the field acting on electrons.

2. The arrangement of nuclei in space is strictly periodical, i.e., the crystalline lattice is assumed ideal.
3. The interaction of electrons is interchanged by an effective field; in other words, the electrons are considered independent and their motion proceeds in an effective self-coordinated field produced by nuclei and electrons.

It should be borne in mind that each electron interacts not only with its own nucleus but also with the atomic frames of the lattice. Within the framework of the given approximation, the task is reduced to the motion of one electron (a monoelectron model) in the defined external field. If the potential energy of an electron in the field is denoted through $U(\vec{r})$, $(\vec{r})$ is the radius-vector of the given point of the space, then it can be easily seen that the relationship $U(\vec{r}) = U(\vec{r} + \vec{\alpha})$ holds true. Here, $\vec{a}$ is the period of the lattice.

To describe the electronic structure of a solid, it is necessary to compare two types of electrons: the inner, which are entirely localized on their nuclei, and the valence electrons, providing a solid conductivity. Conventionally, to describe the energy spectrum of the electronic structure of a solid and kinetic phenomenon two models are utilized. In the first one, the wave function of the valence electron is constructed as a linear combination of atomic wave functions (a hard bond method), and in the second model, in the form of a linear combination of the wave functions of a free electron (a model of free electrons).

Let us consider a one-dimensional infinite chain of common, periodically located monovalent atoms. Generalization for the case of a large number of measurements and multi-valent atoms is not the principal problem that leads to complicated calculations.

The Schrodinger equation of the system considered has the form:

$$-\frac{\hbar^2}{2m_0}\nabla^2\Psi(x) + U(x)\Psi(x) = E\Psi(x). \tag{1.1}$$

Here, m_0 is the mass of the free electron; $\hbar$ is the Plank constant; $\Psi(x)$ is the wave function; E is the proper value of the energy, and x is the coordinate of the electron.

Equation (1.1) explains electron behavior in the crystal for both models. In the first case, the potential $U(x)$ is equal to the sum of atomic (nuclei) potentials composing the lattice of a solid. In the second case, this potential is constant. As mentioned above, the wave function should be sought in the form of a linear combination:

$$\Psi(x) = \sum_g \alpha_g \varphi_g, \tag{1.2}$$

where in the first case φ_g is the wave function of a valence electron in the isolated atom under the number "g", and in the approximation of the nearly free electrons φ_g is the plain wave.

Taking into account that in the periodic potential the wave functions of an electron must satisfy the Bloch theorem, and confining our discussion to the overlapping of the wave functions relating to the neighbouring atoms, we obtain for the electron energy

$$E_n(k) = E_n^0 + \frac{\delta E}{2}\cos(ka). \tag{1.3}$$

Here, parameters E_n^0 and δE are expressed through the proper values of Schrodinger equations for separate atoms, the matrix elements, and the integrals of the overlapping of wave functions; n enumerates the states of atoms; and k is the wave number.

Thus, due to the overlapping of the wave functions of neighbouring atoms and the periodicity of the crystal, a splitting of atomic levels takes place. The atomic level is likely to erode and transforms to a band (δE is the width of the band). Each atomic level corresponds to an allowed band. In addition, the center of the band is shifted with respect to the position of the relevant level. If the function $\cos(ka)$ from Equation (1.3) is expanded in series at small k, then an expression is obtained which has a term formally coinciding with the kinetic energy of a particle (it should be borne in mind that in the periodic potential the electron momentum (k) does not remain and therefore a quasi-momentum is talked about here). Hence, the effective electron mass in the band of conductivity may be introduced:

$$m_e^{*-1} = \frac{\delta E}{2\hbar^2}\alpha^2.$$

It is easy to see that with increasing allowed bandwidth, the effective mass decreases. An expression for the effective electron mass can be written in another fashion. From Equation (1.3) it follows that

$$m_e^{*-1} = \frac{1}{\hbar}\frac{d^2E_n(k)}{dk^2}.$$

The effective mass can be both effective and negative. In fact, for the electron which is found near the ceiling of the allowed band, the quantity $\frac{d^2E}{dk^2} < 0$. In the uniform electrical field, the behavior of such an electron is associated with a quasi-particle, a positive charge, and positive effective mass. Hence in the almost filled-up band of conductivity, when only electrons found near the band ceiling make a non-zero contribution to a current transport, it is convenient to introduce the notion of a hole. Indeed, the transfer of charge and energy corresponds to a motion of positively charged "particles", the number of which corresponds to the number of electrons required to fill out conductivity bands. Such "particles" are called "holes" since their origin is associated with the unpopulated states in the band.

Thus, the electron properties of ideal crystals within the framework of a band approximation are characterized by alternation of allowed and forbidden energy bands. The upper energy band is called a band of conductivity; the lower populated band separated from the upper one by an energy gap is called a valence band. The number of carriers in the conductivity band and their distribution over energies are subjected to the Fermi-Dirack statistics and depend on the position of the Fermi level. The Fermi level is defined as the energy of a state whose probability to be populated is equal to 0.5. The states below the Fermi level are populated. If the conductivity band is partially populated, then the given solid is metal, and in this case an arbitrarily weak external field generates an electrical current. In the case where the conductivity band is vacuous, then, in order to generate a current, it is necessary that a fraction of electrons from the valence band be transferred in the conductivity band by means of an additional energy (thermal motion, light, etc.). In this case, at a small width of the energy gap ($E_g \approx 2 \div 3$ eV) we deal with semiconductors; at a large energy gap width E_g, with insulators.

Charge carriers — conduction electrons and holes — are the simplest elementary excitations of a solid in one-electron approximations. Apart from this, within the framework of a band approximation a number of multielectron effects can be counted. These effects include collective excitation states of a plasmon character. These quasi-particles (plasmons) are the vibrational quantum of the gas electron density of charge carriers. The frequency of plasmons for conduction electrons in metals according to Bohm and Pines is equal: $p = \left(\frac{4\pi n_e e^2}{m_e^*} \right)^{1/2}$, where n_e is the concentration of electrons, e is the charge of the electron.

In the framework of the band approximation, the interaction between charge carriers is not taken into account. The associated effects are called correlative. The Coulomb interaction of electrons and holes with lattice vibrations (electron-phonon interaction) leads to the origin of another quasi-particle called a polaron (this term was suggested by Pekar). The interaction of the charge with ions of the mechanically polarizing medium leads to the appearance of polarization. The potential energy of the charge decreases in the electrical field associated with this polarization. A polarized charge-created potential well arises. The charge transfer over the lattice is accompanied by a displacement of the potential well. The electrons and holes together with the polarized wells created by them are called polarons. If the radius of a potential well essentially exceeds the lattice constant, then we talk about polarons of a large radius.

Account of the Coulomb interaction of conduction electrons and holes allows the notion of a neutral quasi-particle, exciton (to be introduced). Having generated an exciton, the electron and hole move over the crystal as a unit. For crystals with a large dielectric permeability and a small width of the energy gap, the exciton can be presented in the form of a single-electron atom. Then the exciton energy spectrum is characterized by a hydrogenlike spectrum and described by the following formula:

$$E_1 = E_g - \frac{m_{ex}^* e^4}{2\varepsilon^2 \hbar^2 l^2},$$

where m^*_{ex} is the effective mass of the exciton $\left(\frac{1}{m^*_{ex}} = \frac{1}{m^*_e} + \frac{1}{m^*_p}\right)$; m^*_e and m^*_p are the effective masses of the electron and hole; ε is the dielectric permeability of the crystal; l is the level number of the exciton energy.

The exciton is characterized by the radius $a_{ex} = \varepsilon\hbar^2/m^*_{ex}e^2$. If its value is much larger than the lattice constant, then such excitons are called the excitons of a large radius, or the Wannier-Mott excitons. It should be noted that the exciton levels lie in the energy gap, i.e., below the conduction band.

If the radius of the exciton is of the order of the lattice constant, then such excitons are called the Frenkel excitons. They are characteristic of molecular crystals. These elementary excitations of solids correspond to electron-excited states of atoms and molecules. Respectively, allowance is made for singlet and triplet excitons.

When the exciton-phonon interaction is large, the exciton in its motion "drags" behind itself the potential well of atomic displacements and, as in the case of a polaron, becomes incoherent. The motion of such an exciton is not of a wave character but of random walks.

It was found to be convenient to consider the elementary excitation of a solid as a quasi-particle and its population as a gas of quasi-particles. Some quasi-particles can be correlated with the excitations of isolated atoms and molecules; the others are characteristic only of solids. Such analogies are given in Table 1.1.

Table 1.1 Elementary Excitation in Gases and Solids

Processes in atoms, molecules, gases	Quasi-particles in solids
Relative motion of particles	Phonons
Ionization	Electrons and holes
Electron excitation of atoms and molecules	Excitons
	Polarons
	Plasmons

It should be noted that in considering amorphous solids, which most often are a subject of studies in radiation chemistry of solids, it is necessary to utilize the concepts of disordered solid physics. In such systems, the notion of energy bands and quasi-particles is retained since they are determined by the short-range order of the lattice, but there are principle differences, too. In disordered solids, the notion of a quasi-momentum loses its meaning; therefore, to describe characteristics of a solid medium, a notion of coordinates is used. Instead of a forbidden band, a notion of energy gap — domains of forbidden energies — is introduced. There are differences between them since the boundaries of energy bands in disordered systems turn out to be spread. Experimentally it manifests itself as a shift of the absorption band edge to a long wave region as compared to a crystal. The considered quasi-particles (electrons, holes, excitons, etc.) are the elementary excitations in disordered solids. The absence of long-range order or an arbitrary character of the potential field in the disordered structure leads to the so-called Anderson localization of states.

1.2 THE PHONON SPECTRUM

Another property of a solid lattice which is extremely important in chemical processes is the ordering of the motions of molecules forming a medium. The potential barrier, determining the temperature dependence of the chemical reaction rate constant, is affected by the mutual position of reagents. As a result of intermolecular vibrations, the distance between reagents changes and noticeably manifests itself through dependence of particle transfer rate on temperature, properties of the medium and reagents, and the phase state of the substance. These questions are treated in detail in Chapters 4 to 6; here we shall dwell only briefly on the vibrational properties of the ideal lattice.

The simplest model of a lattice retaining the properties of interest to us is a one-dimensional chain of identical interacting atoms. This model allows the idea of the spectrum and dynamic oscillations, on which a relative displacement of particles participating in the chemical reaction may be obtained. In a harmonic approximation, the equations of the motion of interacting atoms are solved easily. By introducing cyclic boundary conditions, we obtain the solution as a sum of harmonic functions with frequencies

$$\omega_q^2 = 4\frac{\gamma}{M}\sin^2\left(\frac{qa}{2}\right). \tag{1.4}$$

Here, γ is the force constant describing atom-to-atom interaction whose mass is equal to M; a is the distance between atoms (the constant of the lattice). The wave number q lies in the limits $\left(-\frac{\pi}{a}, \frac{\pi}{a}\right)$. The allowable values of the frequency ω_q are determined by ratio (1.4) and are in the band $\left(0.2\sqrt{\frac{\gamma}{M}}\right)$. The corresponding quasi-particles characterized by the quasi-momentum $\hbar q$ and the energy $\hbar\omega_q$ are called phonons. The analog of phonons in gases is a relative motion of molecules (see Table 1).

If the lattice is not monoatomic and contains two atoms in the elementary cell, the vibrations of such a system correspond to two dispersion relationships of the type in Equation (1.4). Two phonon branches correspond to them:

acoustic

$$0 \le \omega_q \le \left(\frac{2\gamma}{M_1}\right)^{1/2} \quad \text{at } M_1 > M_2,$$

and optical

$$\left(\frac{2\gamma}{M_2}\right)^{1/2} \le \omega_q \le \left[2\gamma\left(\frac{1}{M_1}+\frac{1}{M_2}\right)\right]^{1/2}.$$

Here M_1 and M_2 are the masses of atoms in the elementary cell. The acoustic branch corresponds to the vibration when the atoms in the elementary cell move in one direction; optical, toward each other.

In the three-dimensional case, new solutions do not appear. We shall note only that in different phenomena some properties and branches of phonons develop. In particular, in order to account for the processes of transfer and dispersion which provide the phonon heat conductivity of crystals, it is necessary to go beyond the framework of the harmonic approximation.

1.3 TRANSFER PROCESSES IN A SOLID

The elementary excitations arising under the effects of radiation lead to various chemical transformations of substance. In molecular crystals, polymers, organic, and many inorganic amorphous substances, due to relatively weak intermolecular interaction molecules, retain their specificity. This, however, does not mean that it is possible to talk about ionization and excitation of isolated molecules, since in the solid phase transfer processes determined by the properties of a molecular lattice play an essential role.

In crystals, the transfer of energy or charge is performed by means of excitons, holes, and electrons. In the amorphous phase, in contrast to the crystal phase the energy transfer takes place during the transfer of electron excitation from molecule to molecule. The propagation of excitation is of a diffused (incoherent) character. The same principle relates to a charge transfer which in amorphous solids also occurs in a diffused fashion. In the crystalline phase these processes may proceed coherently.

Elementary excitations are captured by the acceptors of energy and charge. Depending on the nature of the acceptor, capture processes cause both stimulating and inhibiting (stabilizing) effects on development of the radiation-chemical process. In the first case, the acceptor, chemically less stable than molecules of the basic substance, having received the energy becomes chemically active and enters into a reaction with the molecules, facilitating the process of substance decomposition. In the second case, the acceptor either facilitates the dissipation of energy or undergoes a monomolecular transformation as a result of which stable products are produced.

Thus, the transfers of energy and charge influence the local concentration of chemically active particles, the process of excitation energy dissipation, and, consequently, the radiation-chemical stability of the substance. If chemical transformations manage to happen before the transfer of excitation to the acceptor, then the chemical processes in the solid phase can be considered as reactions of separate molecules interacting with the medium.

Mechanisms of intra- and intermolecular energy transfer will be discussed below. As for a charge transfer, it is treated in detail in monographs[18,19] and there is no need to discuss this issue here. In Sections 3.4 and 7.3, chemical effects of such a transfer and the process of thermalization will be explained briefly.

1.3.1 The Intermolecular Transfer of Energy

Let us consider some mechanisms of a radiationless intermolecular transfer of electron excitation energy in solids. The main transfer mechanisms are induction-resonance, exchange-resonance, and exciton mechanisms. The physical cause of the energy transfer is the interaction of the excited molecule of the donor and nonexcited molecule of the acceptor. When we have an interaction of dipole points of transition of molecules, we talk about the induction-resonance transfer mechanism. Its classical model is a transfer of excitation energy in the system of two mechanically connected oscillators.

According to Förster, in dipole interaction of the donor and acceptor, the probability of transfer is inversely proportional to the sixth degree of the distance between them.[20] The experimental values of the distance, when the probability of the dipole-to-dipole transfer of energy (singlet-to-singlet transfer of energy) is high, are on the order of 50 Å.[21]

The conditions of weak molecular interaction (~10^{-3} eV) and fast vibrational relaxation of the energy of acceptor excitation to prevent the inverse process are the criteria of applicability of the Förster theory. The generalization of the Förster theory for the case of multipole and exchange interactions was conducted by Dexter.[22] For multipole interactions, the degree of the exponent at distances between the donor and acceptor naturally increases.

The exchange-resonance mechanism of the energy transfer is determined by the electron exchange interaction of the donor and acceptor molecules. The transfer probability caused by exchange interaction exponentially depends on the distance between the donor and acceptor and is connected with the overlapping of their wave functions. The characteristic experimental values of the distance for energy transfer caused by the exchange interaction (triplet-to-triplet transfer) is of the order of 10 Å.[21]

The exciton mechanism of the energy transfer is realized in the case when the energy of molecule interaction is sufficient (> 10^{-2} eV) and the relaxation time of excitation energy is larger than the time of energy transfer. The exciton propagation can be presented as the energy transfer over the electron excitation levels of interacting molecules, with the electron level not managing to be localized on one of these molecules.

The exciton mobility and its capture probability by trap are the main characteristics of energy transfer through the exciton mechanism. The exciton mobility can be characterized by the value of diffusion displacement, $L = \sqrt{D\tau_0}$, where D is the diffusion coefficient of the exciton and τ_0 is its lifetime. In case of a weak exciton-phonon interaction (coherent excitons), the coefficient of exciton diffusion can be determined from the Boltzmann equation. Using this equation, a distribution of excitons by states is described both accounting for and not accounting for exciton-phonon interaction. For excitons of a large radius (the Wanier-Mott exciton), the estimations first were made in Reference 23, and for excitons of a small radius (the Frenkel excitons) in Reference 24.

Under a strong exciton-phonon interaction (incoherent excitons), the estimations for excitons of a small radius are carried out in Reference 25. In this case, the motion of excitons is of a random hopping character from molecule to molecule, and the kinetics of the process with participation of excitons is described by a diffusion equation derived on the basis of the model of random walk-outs.

The diffusion coefficients of large radius excitons are of the order of magnitude of 10^2 cm^2/sec and their lifetime is 10^{-6} sec; therefore, the length of the diffusion displacement is of the order of the magnitude of 10^{-2} cm. For singlet excitons of a small radius $D \approx 10^{-3}$ cm^2/sec, $\tau_0 \approx 10^{-8}$ sec, and $L \approx 100$ Å, for triplet excitons, $D \approx 10^{-4}$ cm2/sec, $\tau_0 \approx 10^{-3}$ sec, $L \approx 10^4$ A.

The temperature dependence of the diffusion coefficient of coherent and incoherent excitons is different. For coherent excitons: $D \sim 1/\sqrt{T}$, and for incoherent excitons $D \approx \exp(-E/k_BT)$, where k_B is the Boltzmann constant, E is the activation energy of the transfer process.

Now we consider the capture of excitons by admixture. The rate constant of exciton capture by acceptor depends on the distance between them and decreases rapidly with its increase. Under multiplet interaction, the decrease follows the degree law, and under exchange interaction, the exponential. Such a strong dependence of the probability of exciton absorption on the distance to the acceptor justifies the utilization of the model of the absolutely black sphere for describing the absorption of excitons by acceptors. According to this model, the exciton with a probability equal to the unit is captured by acceptor if the exciton-acceptor distance does not exceed the capture radius characteristic of the given pair.

The absorption rate constant of coherent excitons, in the model of the absolutely black sphere, has the form:

$$K = 4\pi D_0 r \qquad r >> l \tag{1.5a}$$

$$K = \pi r^2 \nu \qquad r << l \tag{1.5b}$$

where D_0 is the diffusion coefficient of the coherent excitons, r is the capture radius, l is the length of the free path of the exciton, v is the exciton velocity. The ratio (1.5a) holds true for a sufficiently large observation time. For incoherent excitons, the capture rate constant can also be written in the form of Equation (1.5a) where D_0 is the diffusion coefficient of the exciton in the model of random walk-outs, and r is approximately equal to the distance between molecules. A detailed description of the regularities of the radiationless energy transfer of electron excitation in solids is given in Reference 26.

1.3.2 Radiationless Conversion and Excitation of Triplet States

In considering intramolecular transitions, one should compare the processes of internal and intercombination conversions.[27,28] The internal conversion corresponds to transitions between states of common multipletness. This is a very fast process; its rate constant is up to 10^{11} to 10^{12} sec^{-1}. In the case when the multipletness of states changes, we deal with the intercombination conversion. Most noticeably, the intercombination conversion happens from the lower excitation singlet level. This is explained by the fact that a fast internal conversion takes place from the upper singlet levels. Therefore, it is easier for the process of intercombination conversion to compete with a radiation transition from the lower excitation singlet state to the ground $k_{ss0} = 10^8$ to 10^9 sec^{-1}. The effectiveness of the intercombination conversion can be characterized by the ratio k_{ST}/k_{ss0} (k_{ST} is the rate constant of the intercombination conversion). The values of this magnitude for a number of substances allow the expectation that the resulting formation of triplet states from the singlet states will be equal to the order of the unit.

The yield of photo- and radiation-chemical reactions depends very much on a molecule lifetime in the excitation state. Especially large is the role of molecules excited to the lowest triplet level when the ground state is singlet (see Chapter 7). The lifetime of excited molecules in the given case is rather large, and they can serve as rather effective sensibilizators.

The properties of molecules excited to the lowest triplet level and the techniques of their registration are considered in References 29 and 30. There are three possible mechanisms of molecule formation in the lowerest triplet state: (1) excitation by electron impact, (2) intercombination conversion, (3) recombination of charges.

Several original papers, reviews, and monographs (see, e.g., References 31 and 32) are devoted to the investigation of inelastic collisions of electrons with atoms and simple molecules. Judging by the data given in these works, it follows that the electron impact transitions with a change of multipletness are possible. The maximum value of the collision cross-section will be somewhat less than in the case of the allowed transitions.[31] In addition, the cross-section value will be essential only in a very narrow range of electron energies near the excitation threshold. It is easy to understand: the excitation of atoms by electron impact with the change of multipletness is only possible due to the exchange interaction of the incident and bond electrons. Such an interaction takes place effectively if the absolute electron energy values will not differ too much.

The excitation of multiatom molecules by electron impact has been studied significantly less. Aromatic compounds that were studied include benzene and naphthalene.[33,34] Singlet-triplet transitions were discovered. The data are available on inelastic collisions of electrons with molecules in pyrene, anthracene, and tetracene for vapors and crystals.[35-38] The comparison of these data allows a conclusion to be made that the results of investigating excitations of various molecular levels by electron impact in vapors may be utilized in studying the radiolysis of solids.

More effectively, singlet states are populated under electron excitation. The excitation cross-section of these states has, as a rule, far larger value in the maximum and also remains noticeable for fast neutrons which is especially important in radiolysis. Being in the excited singlet state, a molecule, as a result of intercombination conversion, may transfer to the triplet state.

Moreover, the triplet states can be formed under charge recombination. It happens as a result of the processes: (1) $A^+ + e^- \rightarrow A^T$ and (2) $A^+ + B^- \rightarrow A^T(A) + B(B^T)$. The probability of molecular formation in the triplet state depends on the energy released under the recombination of charges. It is obvious that the energy must suffice to excite the triplet state. Here, the spin state of recombining charged particles plays a substantial role. To form triplet states, these particles must have parallel spins. The contribution of various origins of excited states to the process of radicals' formation has been studied experimentally in Reference 39. It has been shown that under the effect of γ-radiation on toluene, the excitation states are formed due to direct action of radiation and recombination of ions.

1.4 SURFACE PHENOMENA

Earlier we considered ideal crystals of infinite dimensions having no boundaries. Real crystals have disorders of lattice periodicity connected with finite dimensions, i.e., the surface. The disposition of atoms, ions, and molecules, even on the ideal surface of a crystal, differs from the disposition of particles on the corresponding crystallographic plains in the crystal volume. For instance, the inevitable disorder of periodicity of the lattice structure takes place.

This leads to the fact that in the energy gap new values of energy appear which were not contained in the electron energy spectrum of the infinite crystal. The wave functions of the states are localized (exponentially abate) in the boundary surface layer and therefore are called surface states, while the additional energy levels are called surface levels. It was Tamm[40] who first described them. Thus the presence of an ideal surface leads to the origin of surface states in the energy gap of the crystal.

In real conditions, on the surface of a solid there exist adsorbed atoms and molecules, as well as a number of different individual ('biography") defects of the surface (vacancies, interstitial atoms, stairs, etc.). They also violate the idealness of the lattice field and form surface energy levels.

The exchange interaction of electrons with participation of surface and volume levels leads to charging the surface and forming (due to the necessity of satisfying conditions of electron neutrality of the sample) a space charge. A surface charge and near-surface-layer compensation charge are called the layer of the space charge and screen a solid volume from the surface charge field. As a result of formation of this charge (formed by the surface states of all types), the electrical potential in the near-surface layer of the solid changes, and the energy bands twist. Figure 1.1 shows the energy diagram of the real surface of the crystal in cases of positive and negative charging of the surface. At positive charging, the band surfaces twist downward; at the negative, upward. A potential difference between the surface and volume of a solid is called the surface potential, φ_s. In infinite units, it is expressed as $Y_s = \dfrac{e\varphi_s}{k_B T}$.

Depending on the type of surface potential barrier in the near-surface layer, an increase or decrease in the concentration of charge carriers as compared to the volume takes place. At the positive potential, the band surfaces twist downward and the boundary of the conduction band approaches the Fermi level. Therefore, near the surface a layer with a concentration of electrons greater than that in the volume is formed. At the negative potential, the band surfaces twist upward and the concentration of electrons in the near-surface layer decreases compared to the volume due to the increase in distance between the bottom of the conduction band and the Fermi level.

The thermal electron work of electron yield at the negative surface potential increases by the value $e\varphi_s$ and at the positive surface potential decreases by the same value.

The surface twist of bands may also develop under external photoeffect, which is determined by emission of electrons from the solid when the latter interacts with radiation. So, for example, using a special treatment of the surface (by deposing a thin layer of cesium) it is possible to reduce the energy level of the vacuum with regard to the band edges and to prepare a surface with a negative effective relationship to the electron. In this case, the effectiveness of the electron emission from the solid increases.

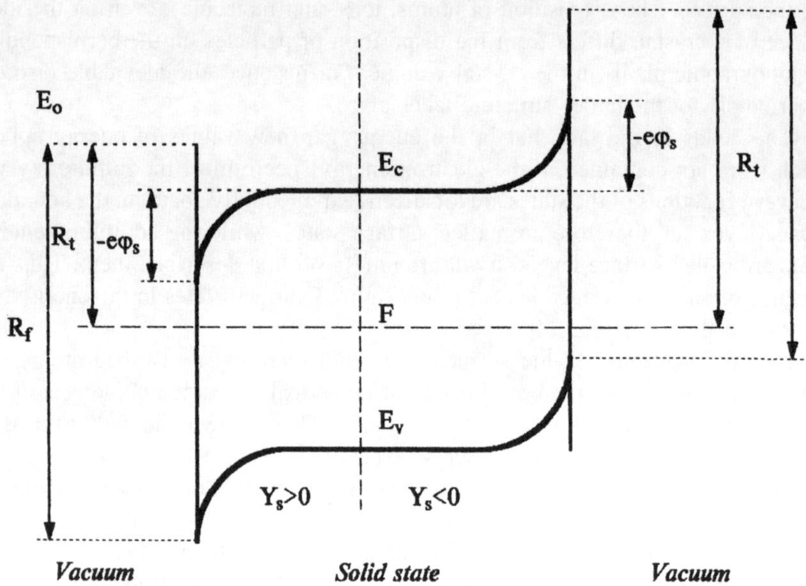

Figure 1.1 Effect of surface charging on electron emission, where E_c = the bottom of the conductivity band; E_n = the top of the valent band; F = Fermi level; E_0 = level of vacuum; R_t = thermoelectron emission; R_f = optic work function.

The surface energy levels also participate in a thermal generation and recombination of electrons and holes. In this connection the lifetime of charge carriers changes. Such influence of the surface is characterized by the rate of surface recombination. The surface recombination rate very strongly depends on the surface potential.[40] In order for the surface level to be a center of recombination, a capture of electrons and holes must take place on it. The electron and hole population of the level depends on its position with respect to the bottom of the conduction band and the boundary of the valence band and is determined by the surface potential. The rate of surface recombination reaches a maximum at a certain value of the surface potential. This value depends on the ratio of the capture coefficients of holes and electrons for the surface level and rapidly diminishes with increasing or decreasing surface potential.

Surface states play a great part in various processes taking place in adsorption and heterogeneous catalysis.[41] Free electrons and holes, being captured to the surface levels of adsorbed molecules, strengthen their bond with the surface. This follows from the fact that the charged molecule, before it transfers to the gaseous phase, needs to transfer an electron or hole to a solid. This process requires some additional energy. The biographic defects of the surface also are the centers of adsorption and, on capturing charge carriers, may adsorb molecules in a charged form. Thus, by changing the concentration of surface defects and the concentration of charge free carriers (e.g., in exposure to light or high-energy radiation), one can change the adsorption properties of the surface. This question will be discussed in Chapter 8.

In the electron theory of hemosorbtion, which plays an important part in describing catalytic phenomena, electrons and holes are considered as free (nonsaturated) valences of the solid surface. Interacting with valence-saturated adsorbed molecules, they convert them into radicals, as a result of which surface chemical reactions become possible.

Moreover, the presence of a solid surface leads to the origin of surface exciton energy levels which are located below the energy band of the volume excitons.[26] Therefore, the states of the surface excitons may play the part of traps for volume excitons and provide the diffusion of excitons directed to the surface. As a consequence, the rate of the exciton surface annihilation increases. Adsorbed atoms (molecules) or contact of a solid with another medium (metal, semiconductor, etc.) are sources of new channels of the exciton's annihilation. The adsorbed atoms (molecules) play the role of exciton traps. Such traps, capturing an exciton, transit into an electron-excitated state or are ionized.

It should be borne in mind that, when in contact with a dielectric with metal or a narrow-band semiconductor, the excitons in the boundary area of the two media may interact with the conduction electrons of metal. Thus, metallic extinguishing of singlet excitons of the molecular crystal due to the transfer of energy to the phase of metal may occur. The energy transfer is accompanied by the birth of surface plasmons and electrons, hole pairs.[42] For triplet excitons, it is more characteristic not to transfer exciton energy to metal but instead to experience the decay of the exciton as a result of which the formed electron or hole is injected into metal.[43] Even a brief description of the surface phenomena shows their importance in radiation-chemical processes in heterogeneous systems.

1.5 DISSOCIATION OF EXCITED MOLECULES

Under the effect of ionizing radiation on a substance, one of the important chemical reactions leading to a formation of radicals is dissociation of excited molecules. Naturally, as a result of the effect of ionizing radiation, not only are excited molecules formed but also excited ions. As is known, the bond energies in molecular ions and corresponding molecules are different which manifests itself in the dissociation rate constant of the excited particle. In addition, the medium also affects the dissociation process; however, in this section we shall not consider the role of environment and bond energy (see Sections 3.1, 6.3, and 7.4). It must be taken into account that the main intermediate products of radiolysis of a number of organic compounds (e.g., hydrocarbons and their derivatives) are atomic hydrogen and corresponding radicals. Therefore, we shall talk here about dissociation with abstraction of a hydrogen atom.

It is necessary to make a distinction between dissociation and predissociation. The dissociation of molecules (radicals, ions) is a process in which the oscillating energy is accumulated in one term, and its amount is sufficient for reaching a state with $E \geq 0$ (Figure 1.2), at which point the reaction products scatter. The source of energy in the given case is not important. This may be a thermal energy, randomly accumulated in one bond, or the energy of high frequency or infra-red laser radiation. Also possible is a transition of electron excitation energy (excitation under the effect of secondary electrons or light) into vibrational, with follow-up fluctuational redistribution and concentration of this energy on the abstraction bond.

If the excitation takes place under the effect of an external source of energy, then the dissociation rate constant may depend in a complicated fashion on the

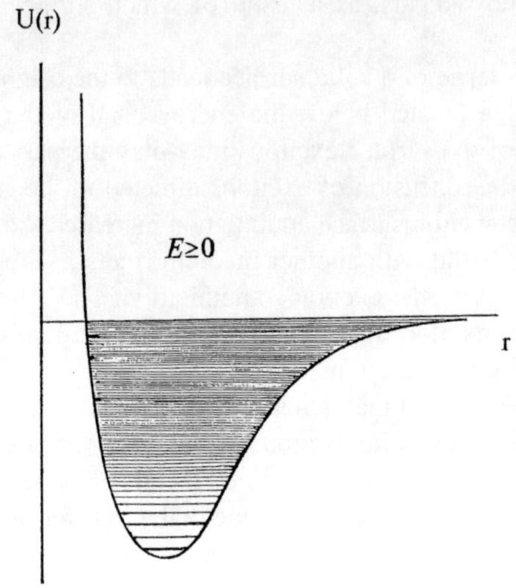

Figure 1.2 Electron therm of molecule. States with $E \geq 0$ correspond to continuous spectrum. At $E < 0$, states of the molecule are bound.

intensity of radiation. In case of thermal activation, the rate constant is of the conventional Arrhenius type:

$$k(T) = \nu \exp[-E_a/k_B T]$$

The activation energy E_a, as a rule, is determined only by the height of the potential barrier in the saddle point. Only in some cases (see Sections 4.1 and 5.4) is an activation energy value influenced by medium properties. As far as the pre-exponential factor is concerned, various approximations are used for its calculation. First of all, we note that, although a dissociation reaction is a monomolecular process but without influence of the environment (in gases, this is collisions with neighbouring molecules), the activation of molecules is impossible. Therefore, calculation of the preexponential factor in many cases takes into account the interaction with neighbouring particles.

A detailed discussion of techniques, allowing a pre-exponential factor to be calculated with some precision, is given in monograph.[14] We shall note only that currently the most promising techniques are a theory of transition state[44] and the Kramers theory.[45] The latter is correct in the assumption that in the selected chemical bond a vibrational energy equal to activation energy is accumulated under interaction with the environment. For the intramolecular potential, a simplest quadratic type is usually chosen. The Kramers approach is applicable for deep potential wells when the energy required to break a chemical bond noticeably exceeds the thermal energy.

Predissociation is the transition of a molecule from an electron-excited state — to which a binding potential corresponds (curve 1, Figure 1.3a and b) — to the region of the continuous spectrum (curve 2, Figure 1.3a and b). Then passing the stage of energy redistribution, the dissociation products scatter. The final term of the molecule can be both binding (Figure 1.3a) and repulsing (Figure 1.3b).

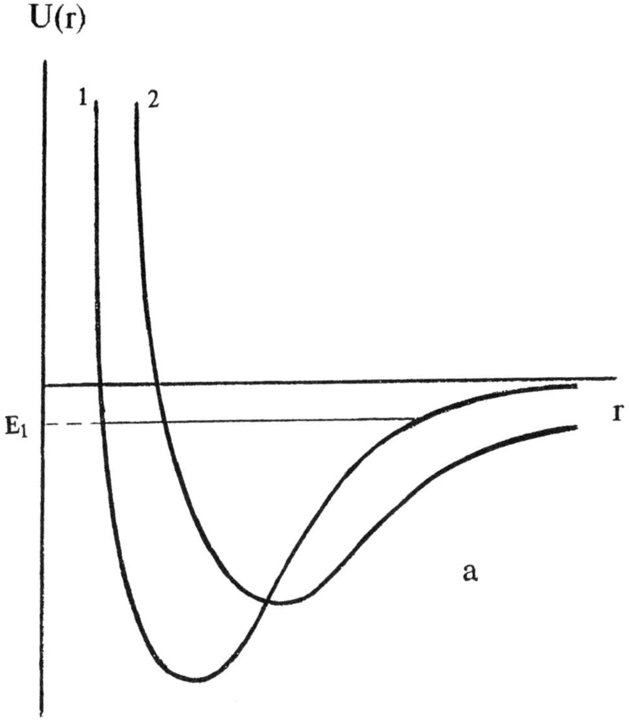

Figure 1.3 Scheme of molecule predissociation. Case (a) corresponds to transition on the bound therm of the molecule. Energy E_1 corresponds to the states of continuous spectrum. Case (b) corresponds to transition on the repulsive therm of the molecule.

The dissociation mechanism is realized only for small molecules, because the probability of energy concentration on a single bond exponentially decreases with an increasing number of bonds. For multi-atomic molecules, the most probable mechanism is the mechanism of predissociation.

A very important question about the influence of the relationship between a binding energy and the position of energy levels on radiation-chemical molecule stability was considered in Reference 46. It was pointed out that the position of the lower singlet levels plays an essential part. If the energy of a lower excitated singlet state is larger than the energy of the breaking bond, then such molecules are unstable in regard to the action of radiation.

As a matter of fact, the defining parameter is the lifetime of a molecule in a reactivity state. For a successful course of chemical processes, the inverse lifetime of this state should be comparable to the dissociation constant rate of a particle studied. In this regard, the position of the lowest triplet level[47,48] plays an important part. The lifetime with regard to deactivation at this level is large, and if its energy exceeds the dissociation potential, then such molecules will be unstable under radiation.

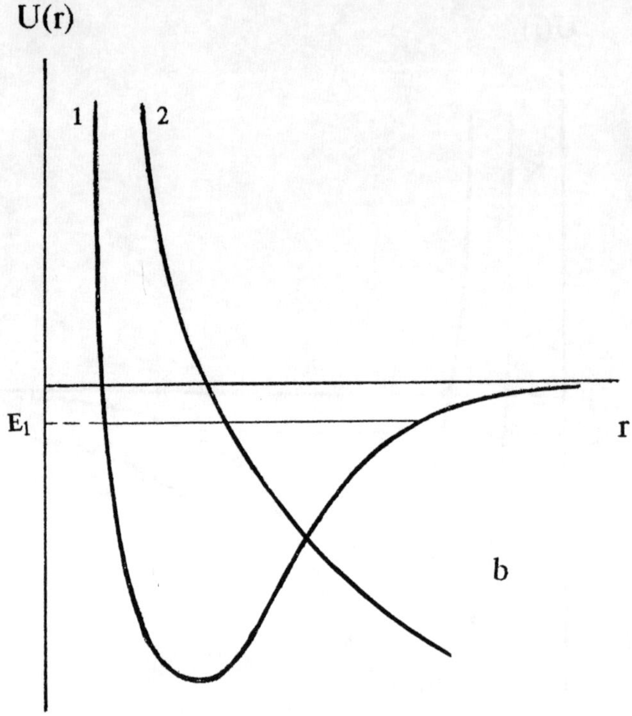

Figure 1.3 (b) (continued)

REFERENCES

1. Kittel, C., *Introduction to Solid State Physics,* John Wiley & Sons, New York, 1956.
2. Kittel, C., *Quantum Theory of Solids,* John Wiley & Sons, New York, 1963.
3. Ziman, J.M., *Principles of the Theory of Solids,* University Press, Cambridge, 1972.
4. Ziman, J.M., *Electrons and Phonons. The Theory of Transports Phenomena in Solids,* Clarendon Press, Oxford, 1960.
5. Pines, D., *Elementary Excitations in Solids,* W. A. Benjamin, New York, 1963.
6. Harrison, W.A., *Solid State Theory,* McGraw-Hill, New York, 1970.
7. Fano, U., Fano, L., *Physics of Atoms and Molecules. An Introduction to the Structure of Matter,* The University of Chicago Press, Chicago, 1979.
8. Calvert, J.G., Pitts, J.N., *Photochemistry,* John Wiley & Sons, New York, 1965.
9. Barltrop, J.A., Coyle, J.D., *Excited States in Organic Chemistry,* John Wiley & Sons, New York, 1975.
10. Turro, N.J., *Modern Molecular Photochemistry,* Benjamin/Cummings, Menlo Park, CA, 1978.
11. de Mayo, P., *Rearrangements in Ground and Excited States,* Vol. 3, Academic Press, New York, 1980.
12. Okabe, H., *Photochemistry of Small Molecules,* John Wiley & Sons, New York, 1980.
13. Nikitin, E.E., Zülicke, L., *Selected Topics of the Theory of Chemical Elementary Processes,* Springer-Verlag, Heidelberg, 1978.
14. Robinson, P.J., Holbrook, K.A., *Unimolecular Reactions,* Wiley-Interscience, London, 1972.
15. Stoneham, A.M., *Theory of Defects of Solids,* Clarendon Press, Oxford, 1975.

16. Morrison, R.S., *The Chemical Physics of Surface,* Plenum Press, New York, 1977.
17. Mott, N.F., Davis, E.A., *Electronic Processes in Non-Crystalline Materials,* Clarendon Press, Oxford, 1971.
18. Ulstrup, J., *Charge Transfer Processes in Condensed Media,* Springer-Verlag, New York, 1979.
19. Khairutdinov, R.F., Zamaraev, K.I., Zhdanov, V.P., *Electron Tunneling in Chemistry,* Elsevier, Amsterdam, 1987.
20. Förster, Th., 10th Spiers Memorial Lecture — Transfer mechanisms of electronic exitation, *Disc. Farad. Soc.,* 27, 7, 1959.
21. *Radiationless Transfer of the Electronic Excited Energy,* Ed., Ermolaev, V. L., Nauka, Leningrad, 1977 (in Russian).
22. Dexter, D.L., A theory of sensitized luminescence in solids, *J. Chem. Phys.,* 21, 836, 1953.
23. Anselm, A.I., Firsov, Yu, A., Length of free path delocalization exciton in polar crystal, *Sov. Phys. JETF,* 30, 719, 1956.
24. Agranovich, V.M., Konobeyev, Yu, V., On the length of the exciton free path in a molecular crystal, *Optics Spectrosc.,* 6, 242, 1959.
25. Trlifai, M., Das Energiespektrum und die Mittlere Freie Bahn der Deformierenden Exitonen in Molekular Kristallen, *Czech, J. of Phys.,* 13, 644, 1963.
26. Agranovich, V.M., Galanin, M.D., *Energy Transfer of Electron Exitation in Condensed Matter,* Nauka, Moscow, 1978 (in Russian).
27. Jortner, J., Rice, S.A., Hochstrasser, R. M., Radiationless transitions in photochemistry, *Adv. Photochem.,* 7, 149, 1969.
28. Advouris, P., Gelbart, W.M., El-Sayed, M.A., Nonradiative, electronic relaxation under collison-free conditions, *Chem. Rev.,* 77, 793, 1977.
29. Singh, A., Triplet state formation in pulse radiolysis, *Rad. Res. Rev.,* 4, 1, 1972.
30. Alfimov, M.V., The triplet states in radiation chemistry, *High Energy Chem.,* 6, 3, 1972.
31. Massey, H.S.W., Burhop, E.H.S., *Electronic and Ionic Impact Phenomena,* Clarendon Press, Oxford, 1952.
32. Mott, N.F., Massey, H.S.W., *The Theory of Atomic Collison,* Clarendon Press, Oxford, 1965.
33. Compton, R.N., Huebner, R.H., Reinhardt, P.W., Christophorou, L.G., Threshold electron impact excitation of atoms and molecules: detection of triplet and temporary negative ion states, *J. Chem. Phys.,* 48, 901, 1968.
34. Doering, J.P., Low-energy electron-impact study of the first, second and third triplet states of benzene, *J. Chem. Phys.,* 51, 2866, 1969.
35. Koch, E.E., Otto A., Characteristic Energy Losses of 30-kev Electron in Vapor of Aromatic Hydrocarbons, *A Acta Opt. Comm.,* 1, 47, 1969.
36. Kunstreich, S., Otto, A., Anistropy of Electron Energy Less Spectra in Athracene single crystals, *A Acta Opt. Comm.,* 1, 45, 1969.
37. Jager, J., Energieverluste monochromatischer electronen in kindensierten aromatischen Kohlenwaserstaffen, *Ann. Phys.,* 22, 147, 1969.
38. Schott, M., Remarks on the process of carrier generation in electron-bombarded crystalline anthracene, *Mol. Crystals,* 5, 229, 1969.
39. Noda, S., Ohi, M., Ershov, B.G., Yoshida, H., Role of ionic processes in the formation of free radicals in g-irradiated glassy toluene, *J. Phys. Chem.,* 84, 2210, 1980.
40. Rzhanov, A.V., *Electronic Processes on Semiconductors Surface,* Nauka, Moscow, 1971 (in Russian).
41. Volcenstein, F.F., *Physical Chemistry of Semiconductors Surface,* Nauka, Moscow, 1973 (in Russian).

42. Kuhn, H., Classical aspects of energy transfer in molecular systems, *J. Chem. Phys.,* 53, 101, 1970.
43. Singh, J., Baessler, H., Theory of excitation dissociation in molecular crystals at the interface of a metal, *Phys. Stat. Sol. B,* 62, 147, 1974.
44. Eyring, H., Lin, S.H., Lin, S.M., *Basic Chemical Kinetics,* John Wiley & Sons, New York, 1983.
45. Krames, H.A., Brownian motion in a field of force and the diffusion model of chemical reactions, *Physica,* 7, 284, 1940.
46. Voevodskii, V.V., Molin, Yu, N., On the radiation stability of solid organic compounds, *Rad. Res.,* 17, 366, 1962.
47. Ovchinnikov, A.A., Plotnikov, V.G., Non-adiabatic reaction of hydrogen atom cleavage from electronically excited molecular system, *Chem. Phys.,* 21, 349, 1977.
48. Plotnikov, V.G., Ovchinnikov, A.A., Photo- and radiation-chemical molecule stability. Monomolecular reaction of hydrogen atom obstruction, *Russ. Chem. Rev.,* 47, 444, 1978.

2 INTERACTION OF RADIATION WITH A SOLID

A radiation-chemical process starts when a substance begins to absorb high-energy radiation. A characteristic feature of the first stage is a spatial nonuniformity of energy absorption. This is an extremely important property which greatly determines kinetics of the following stages in radiolysis, including the chemical ones. The kinetics of many stages of a radiation-chemical process are diffusion controlled. At the same time, if a reaction is controlled by diffusion, the dependence of an active particle concentration on time does not obey simple laws of the first and second orders. In this case, the kinetics depend on the initial spatial distribution of active particles.

Highly specific features develop in the course of heterophase radiation-chemical reactions. It is of interest that the energy is released in one phase and then, by means of electrons and excitons, is transported to another phase where chemical conversions happen. The transfer of energy and charge is provided by short-lived atomic excitations of a solid. These excitations are chemically active and their reactivity is determined by medium properties.

2.1 ABSORPTION OF IONIZING RADIATION BY A SOLID

Interaction of high-energy radiation with atoms and molecules of the irradiated substance gives rise to a radiation energy transfer to the medium. The regularities of primary processes occurring when a solid absorbs radiation differ little from those in gases and liquids. The primary processes include: formation of electron-positron pairs, Compton scattering, photoeffect, excitation, and ionization with fast electrons and ions as a result of inelastic collisions. When high-energy radiation is absorbed, the inner-shell electrons of atoms and molecules (whose energy do not depend on the aggregate state of a substance) play an important role. This is why the differences in ionization potentials for substances in the solid phase, liquids, and gases do not influence interaction processes of high-energy radiation with a substance.

Let us consider the basic physico-chemical processes which occur under the radiolysis of solids. At the physical stage, under interaction of high-energy radiation with a substance (γ-radiation, X-ray radiation, charged particles), fast electrons are formed.[1] The simplest process which results in the formation of electrons is photoelectric absorption (a photoeffect). It should be borne in mind that in this case we talk about the internal photoeffect under which the photon energy is transferred to the inner-shell electron of the atom. The knockout electron energy is determined by the Einstein equation:

$$E = \hbar\nu - E_a,$$

where $h\nu$ is the photon energy and E_a is the electron-binding energy in the atom.

To obtain an absorption coefficient, it is necessary to know the wave functions of the knockon electron in the initial and final states. Expressions for wave functions of complex atomic systems may be obtained only approximately. The mass coefficient of photon absorption due to the photoeffect has the form:[2]

$$\tau \approx \left(\frac{Z}{\hbar\nu}\right)^3 (1 + 0.008\,Z),$$

where Z is the atomic number of the medium. It is obvious that the energy absorption due to a photoeffect plays a noticeable part in the photon energies $\hbar\nu < 0.1$ MeV and for elements with high atomic (high-Z) numbers.

After having absorbed a high-energy photon by knocking out an inner-shell electron, the inner-shell filling with outer-shell electrons is possible by one of two processes: radiation of the X-ray photon or knocking out one more electron (the Auger effect).

Another mechanism of the energy absorption is Compton scattering, which may be considered as an elastic collision of a photon with a free or weak-bond electron. In this case, the photon loses a fraction of its energy, and, owing to the difference of energies, a recoil electron is knocked out from the atom (Compton electron). The energy of the Compton electron is determined by the following expression:[3]

$$E = \frac{\left(\hbar\nu/m_0c^2\right)(1-\cos\theta)}{1+\left(\hbar\nu/m_0c^2\right)(1-\cos\theta)},$$

where m_0 is the mass of the free electron, c is the light velocity, ν is the frequency of the incident photon, θ is the scattering angle of the photon.

The calculations show that the mass coefficient of absorption changes little with regard to photon energy and the mass coefficient of absorption has the maximum.[4] Its value is about 10^{-1} cm^2 g^{-1}. It should be noted that, for the ordinary γ-emitters used in radiation chemistry, the photon energy is of the order of 1 MeV (Co^{60}, 1.12 and 1.33 MeV, Cs^{137}, 0.66 MeV). For these sources, the Compton effect is the principle mechanism of energy absorption.

Another sufficiently effective process which occurs due to the interaction of a photon with the atomic nucleus is the birth process of electron-positron pairs. To form a pair, it is necessary for the photon energy to exceed the rest energy of the pair equal to $2m_0c^2$ (1.02 MeV). After annihilation of the pair, two photons of 0.51 MeV are formed. This mechanism makes an important contribution to energy absorption only if the photon energy is larger than several MeV.[4]

A further absorption of energy occurs as a result of the interaction of the fast knockout electrons with the medium. Let us consider the basic regularities of a charged particle interaction with atoms and molecules. As a result of such an interaction, the charged particle loses its energy and the atom or molecule changes its electronic state, the particle becoming ionized or electron-excited. If the energy transferred is much more than the potential of the ionization, the knockon electron may be treated as a free one. In this case, the scattering cross-section is described by a well-known Rutherford formula. Such collisions are called knock-out collisions. The theory of collisions is treated in detail in monographs.[5-7]

For distant or glancing collisions resulting in excitation of atoms, the cross-section is estimated by the Bethe formula.[8] For nonrelativistic electrons ($\nu/c \ll 1$), the mean ionization losses of energy per unit path length are calculated by the formula:

$$-\frac{dE}{dx} = \frac{4\pi e^4 n_0 Z}{m_0 \nu^2} \ln \frac{m_0 \nu^2}{I},$$

where m_0 is the electron mass, n_0 is the density of the medium particle, and I is the effective potential of the medium excitation. The value I is usually determined experimentally. In addition, different methods of its calculation are available.[9]

Since the energies of incident charged particles are not far in excess of the electron binding energy in the atom, it is necessary to introduce a shell correction for the electron-binding effect. In this case, the loss of the electron energy is expressed as follows:

$$-\frac{dE}{dx} = \frac{4\pi e^4 n_0 Z}{m_0 \nu^2} \ln \left[\frac{m_0 \nu^2}{I} - \frac{C}{Z} \right],$$

where C is the shell correction constant allowing an electron binding energy to be determined. More details are given in References 10 and 11.

This correction holds true for media with low atomic (low Z) numbers. Relativistic corrections to the Bethe formula at high velocities of charged particles can also be determined.[12]

In the case of molecular media, the Bragg rule[8] is utilized for determining the atomic number Z and effective potential of ionization I of the medium. To determine these values, the following expressions are used:

$$Z = \sum_a n_a Z_a \text{ and } Z \ln I = \sum_a n_a I_c,$$

where n_a is the number of atoms of the kind "a" in a molecule, Z_a is the atomic number of the atom of the kind "a", I_a is the ionization potential of the atom of the kind "a".

Until now we have talked of the energy loss mechanism during the interaction of high-energy radiation with a substance. Now we consider the spectrum of high-excitation states formed in the irradiated medium under whose relaxation chemically active elementary excitations of solids are brought about. These include states arising as a result of a cascade of ionization by means of "shaking-up" and Auger-effect and, in addition, "superexcited" states and collective states of the predissociation type.

An ion with an inner-shell vacancy possesses a high excitation energy, which, as a rule, is transferred to the other electrons of the molecule, which leads to an extra ionization. In particular this energy may be transferred to one or several electrons of the outer orbitals (the Auger-effect[13]). The extra ionization is also caused by a sharp change of the electromagnetic field in which the outer-shell electrons move. Such a change occurs when an inner-shell electron is knocked out (a "shake-up" mechanism).[14]

As a result of the above mechanisms, multicharged ions are generated. These ions are extremely unstable and rapidly disintegrate into several positive fragments. The probability of a multiple ionization is very high. So, for example, in the case of Ne knocking out an electron from the K-shell, the probability makes up 99% and from the L-shell, it makes up 13%.[15]

In the condensed phase, the energy absorption by a charged particle may take place with the creation of collective excitations of the plasmon type.[16] In metals, as is known, the plasma oscillations of conduction electrons are activated. At the same time, the molecular media are, as a rule, isolators; i.e., the conduction band has no electrons. In this case, under absorption of charged particle energy, the plasma oscillations of the valence electrons are activated, their frequency being determined by the density of the valence electrons. An approximate formula for determining the plasma frequency is described as follows:[17]

$$\hbar w_p = 28.8\ (N_v\ \rho/M)^{1/2},\ eV$$

where N_v is the number of the valence electrons per molecule, M is the molecular weight, and ρ is the density of the medium in g/cm^3.

Such excited collective states are responsible for wide, intensive maxima in the range of 15 to 25 eV in the energy loss spectra of fast electrons. A detailed description of the properties of plasma oscillations is given in References 18 and 20.

Molecules composing a medium may be in an excitation state, the energy of which is much larger (by 10 eV and more) than the first potential of a molecule ionization. The lifetime of such states is 10^{-11} to 10^{-14} sec. Their decay, as a rule, occurs due to autoionization. These states of a molecule were named by Platzmann as "superexcitation" states.[21] In the literature, the following ways of forming super-excitation states of molecules are considered:

1. Simultaneous excitation of two valence electrons
2. Population of the high-excitation Ridbergian states of molecules
3. Excitation of an inner-shell electron to a free orbital

An experimental study of the given phenomena has shown that the first items on the list of possible means of formation of superexcitation sates of molecules are realized.[22] A radiation yield of superexcitation states is estimated to be 0.6 to 1.[23] The decay may happen due to autoionization, predissociation, radiation, or radiationless relaxation. A description of the processes with participation of superexcitation states is presented in References 18, 19, and 24. The most noticeable differences in the radiation effects of a substance in various aggregate states arise as a result of elastic collisions of heavy bombarding particles (protons, neutrons, α-particles, heavy ions) with medium molecules.

The "biography" defects, whose origins are connected with the formation conditions of the solid phase, and radiation-induced defects influence the processes of formation and transfer of chemically active elementary excitations. Apart from elastic collisions, radiation defects are formed effectively under inelastic impact, too. These processes are primarily under the effect of γ-quanta and fast electrons on substances.

The most frequently encountered defects which appear under elastic interactions are the Frenkel defects (a vacancy and nearby interstitial atom). These defects are formed if the transferred energy is enough to move a molecule (atom) to the interstice. In a case where the amount of the transferred energy is sufficient, the initially displaced atom is able to cause a cascade of secondary and further displacements. As a result, cascades of vacancies and interstitial molecules (atoms) are formed in a solid.

When, under collisions, such an amount of energy is transferred that the free path length of a knockon particle sufficiently exceeds the constant of the crystal lattice, the solid structure does not cause an essential influence on the development of the cascade of displacements. The process takes place in the same way as in amorphous bodies. If the energy transferred decreases, then the structure of a crystal lattice begins to manifest itself. Displacements chiefly occur toward the most dense packing. One should note also the possibility of the formation of a sequence of collision focusings which transfer energy over a chain of atoms (focusons).[25] Excitation of focusons that may disperse their energy on thermal oscillations leads to a decrease in the number of displacements in the cascade.

Under collisions of heavy particles with the lattice, formation of a dynamic particle (croudion) may occur. A croudion is a condensation of atoms in which $n + 1$ atoms occupy n sites toward the most dense packing. As a result of the sequential displacement collisions of atoms, the croudion transfers over the sample. After completing atomic collisions generating a croudion, the n-positioned atom, which was previously knocked out from the initial position, transfers in the interstice. Effectively this corresponds to transfer of mass and energy. Thus, vacancies and an interstitial atom dispersed at sufficient distances are formed.[26,27] The so-called channeling effect[29] leads to a similar result in freely packed structures. The particle is likely to find itself in a channel in which it undergoes a guiding effect, with the energy losses sharply decreasing and the depth of penetration increasing.

Elastic collisions play an important role in radiation-chemical processes in solids irradiated by heavy ions. A sufficient energy is released in a small volume which leads to local meltings and causes temperature-activated chemical transformations of substance. Such collisions are important under neutron irradiation of a substance. In this case, the most probable primary processes are elastic collisions of neutrons with nuclei. This results in the appearance of high-energy ions which produce the following ionization and excitation. Elastic collisions also should be taken into account under nuclear conversions caused by the effect of various types of radiation. Fission fragments formed in nuclear reactions are capable of producing sufficient destructions in a substance and causing a number of chemical transformations.

The Frenkel defects are very often a result of ionizing radiation effect on a substance (the neighbouring vacancies and interstitial atom). So, in ionic crystals the multiple ionization of a negative ion is possible (i.e., its conversion into a positive ion). The resultant positive ion can manage to go into the interstice and stabilize until it recombines with an electron.[30] A transfer of energy to the lattice also happens (sufficient to produce a defect) under the localization of a radiation-induced exciton or under the recombination of an electron and hole.[31] Apart from these mechanisms, the formation of Frenkel defects may also be caused by a collapse of high-energy polarons. Such processes are characterized by a reduced energy of defect formation, and the mechanism itself is called a subthreshold formation of defects.

The aspects of defect formation under interaction of high-energy radiation with a solid become more complicated due to formation of complex defects (divacancies, clusters, etc.), annealing of defects due to migration toward the channels (grain boundaries, dislocations, etc.), and chemical transformations of a substance (radicals, stable products of radiolysis) even in the case when the energy of the primarily displaced atom is less than the threshold energy of displacement, i.e., under irradiation by γ-quanta and electrons. One can get acquainted in detail with defect formation under radiation-substance interaction by referring to References 32 and 33.

2.2 SHORT-LIVED, CHEMICALLY ACTIVE, ELEMENTARY EXCITATIONS IN A SOLID

The simplest elementary excitations of a solid arising under exposure to radiation are conduction electrons and holes, as well as excitation states (excitons, excited molecules).

2.2.1 Free Electrons and Holes

As is known, a radiation yield of electrons and holes (ion pairs) at ionizing in the primary processes of energy absorption is determined by the Bethe formula.[8] To use this formula in practice it is necessary to know either a theoretical or experimental value of the ionization cross-section of atoms and molecules. Semi-empirical formula for calculating ionization cross-sections are given in References 34 to 36.

It should be taken into account that for multiatomic molecules the ionization cross-section may be presented in the form of an additive value. This follows from the fact that, for a relativistic electron, the wave length is very small and the scattering cross-section is an additive value over different dissipation centers. The analysis of experimental data by ionization cross-sections of some hydrocarbons has shown that using the given property one may calculate ionization cross-sections of arbitrary molecules with precision to 2%.[37,38] To determine the total yield of electrons and holes, it is necessary to take into account the various processes in which they are formed. This may happen in an elastic scattering of secondary electrons and multiple ionization, as well as a decay of superexcitation states and plasmons.

For gases the energy of ion pair creation is close to 25 eV, which gives the value of a radiation yield of the pair equal to ~4.[39] In the condensed phase, the radiation yield is somewhat higher. So, for liquid argon, krypton, and xenon, the radiation yield values become 4.5, 6.0, and 6.6, respectively.[40] This is larger than the radiation yield in gases of the values of 0.7, 1.9, and 2.1. A total radiation yield of electrons and holes for some solid organic samples has been determined (e.g., for 3-methylpentane at 77 K it equals 5.4 ± 0.5 1/100 eV).[41] Such a difference is usually connected with a decrease in the potential of atoms and molecule ionization in the condensed phase in comparison with the gaseous phase. In fact, as mentioned in Section 1.1, the energy of the forbidden energy band (corresponding to the potential of ionization in gases) is 1 to 3 eV less than the ionization potential of atoms and molecules in the gaseous phase. Thus, to create a pair in the solid phase, less energy is required.

A decay of collective excitations of a solid — plasmons — is another channel for formation of electrons and holes. The absorption of a plasmon as a result of the electron-plasmon interaction is virtually identical to the absorption of a photon with the same energy. The collective excitation in this case transforms into a one-particle. The lifetime of a plasmon state is estimated to be 10^{-16} sec.[42]

In irradiating a substance with ionizing radiation, a noticeable number of electrons is created whose energy is insufficient for exciting electron states. These are so-called subexcitation electrons.[43] The process of the energy loss by such electrons, up to energies on the order of k_BT, is called thermalization. The subexcitation electrons may spend energy to excite phonons and internal vibrations in molecules and may be autolocalized. In addition, they are captured by neutral molecules and ions. In polar substances, a loss of energy by electrons in dipole relaxation of a molecule takes place. More details about the mechanism of the energy loss by subexcitation electrons can be found in Reference 20.

The length (l_{th}) and the time of thermalization (t_{th}) are important characteristics of the processes of subexcitation electron deceleration. They depend very strongly on a medium polarity. So, for the polar media of the water type l_{th} ~ 2.0 nm, t_{th} ~ 10^{-14} sec^{-1}; for hydrocarbons l_{th} ~ 5.0 to 20.0 nm, t_{th} ~ 10^{-12} sec^{-1}; for inert gases where electrons lose energy only in elastic collisions, l_{th} ~ 100 nm, t_{th} ~ 10^{-9} sec^{-1} (see, for example, References 44 and 45). For a polar medium, the thermalization length is small and correlated electron-hole pairs may be considered as exciton or electron-excited states. The thermalized electrons may stay in quasi-free (conduction electrons) and localized states. The ion pairs which avoided a heminal recombination are usually treated as quasi-free carriers. The probability that electrons do not recombine with the parent ion is determined, according to Onzager,[46] by the following expression:

$$P = -e^{-r_c/r}, \quad \text{where } r_c = \frac{e^2}{\varepsilon k_B T}.$$

Here, r is the distance between the electron and ion, e is the dielectric permeability, k_B is the Boltzmann constant, and T is the medium temperature.

The critical radius is determined from the equality condition of the Coulomb interaction energy and thermal energy (k_BT). The average time of a heminal recombination is estimated by means of expression:[47]

$$\tau_r = \frac{<r>^3}{3e\mu},$$

where $<r>$ is the average distance between electron and ion depending on the initial energy of the electron, μ is the mobility of the pair of an electron and positive atom.

A yield of quasi-free pairs depends on the distribution of pairs by distances between electron and ion $f(r)$:[48]

$$G_{quasi-pairs} = G_{total} \int_0^\infty f(r)P(r)dr.$$

For $f(r)$, a choice is often made for the exponential distribution, although other functions are used, too.

The above-mentioned concepts may be applied to a solid in the case when a short-range order is absent, i.e., the exit outside the framework of a band approximation. Such solids include glasses. For substances where a short-range order remains, it is necessary to use notions of solid-state physics. For example, one cannot talk about a heminal recombination in crystals. As mentioned in Chapter 1, the electrons and holes are delocalized, and it is impossible to indicate to what atom they belong.

When a high-energy radiation interacts with a substance, a regeneration of nonequilibrium current carriers — electrons and holes (analogs for which, in gaseous and liquid phases, are subexcitation electrons and ions) — takes place. With time, they disappear as a result of the recombination process. The recombination is characterized by the rate of recombination and the mean lifetime of nonequilibrium current carriers τ. In other words, the rate of recombination is a probability for a pair of excess carriers to disappear per unit time.

Subexcitation electrons are characterized by an energy higher than k_BT and a quasi-momentum different from zero. The ways of spending an excess energy have been described above. Usually, the relaxation time of the momentum is much less than that of the energy τ_E. If all $\tau_E << \tau$, then the electrons manage to be thermalized. At the same time, at $\tau_E \geq \tau$, it is necessary to take into account a change of the electron mobility. The influence of a change in mobility is usually manifested at low temperatures.

The lifetime of nonequilibrium carriers is much influenced by their concentration. Usually a notion of an instant lifetime is introduced, i.e., a lifetime of nonprincipal carriers at the given values of concentration. In organic solids, the rate of the interband recombination due to a large width of the forbidden band is small. Therefore, a recombination with the participation of impurities and defects is the main mechanism of recombination. Details about charge recombination processes may be found in References 49 and 50.

There are some publications reporting studies of the mobility of quasi-free electrons (the conduction electrons) in organic and inorganic solids. As a rule, the mobility is higher in solids than in liquids. Data may be given characterizing the value of the electron mobility in various compounds. For solid argon (82 K), krypton (113 K), and xenon (157 K), the mobility is 1000, 3700, and 4500 $cm^2/V\ sec^{-1}$, respectively, and for the liquid state 475 (85 K), 1300 (120 K), and 2200 (163 K) $cm^2/V\ sec^{-1}$.[51] For ice[52] at 208 K, the mobility of electrons equals 25 $cm^2/V\ sec^{-1}$; for frozen solution, NaOH[53] (1.0 *M*) = 1.8 $cm^2/V\ sec^{-1}$, and for liquid water at 295 K, = $1.8 \cdot 10^{-3}$ $cm^2/V\ sec^{-1}$.[54] In the crystals of aromatic compounds,[55] the mobility of electrons is on the order of 1 $cm^2/V\ sec^{-1}$. In neopentane, the mobility increases from 70 (295 K, liquid)[56] to 170 $cm^2/V\ sec^{-1}$ (295 K, solid body).[57] In liquid alkanes,[58] the mobility of electrons changes within the limits from 10^{-2} to 100 $cm^2/V\ sec^{-1}$. Thus, it may be assumed that for many organic solids where the mobility of conduction electrons is on the order of 1 $cm^2/V\ sec^{-1}$, a band mechanism of the charge motion is realized. Only this mechanism may provide such a high mobility of electrons. In addition, in the given case a mobility must increase with decreasing temperature; this is what happens in a frozen solution NaOH.[53]

At the same time, for 3-methylpentane and cyclohexane much lower values were obtained. For cyclohexane,[59] the mobility of electrons is equal to $7 \cdot 10^{-2}$ $cm^2/V\ sec^{-1}$ at 275 K; for 3-methylpentane,[60] $2 \cdot 10^{-2}$ $cm^2/V\ sec^{-1}$ at 35 K, increasing with increasing temperature. In this case, the motion of electrons seems to occur by a jump mechanism and it must be considered within the framework of disordered solids (see Chapter 1).

2.2.2 The Trapped Electrons and Holes

The regularities of the origin and behavior of trapped electrons e_t^- have been studied rather well.[61,62] Trapped electrons are characterized by wide optical bands of absorption (~1500 to 2000 nm), the maximum absorption (λ_{max}) in the optical spectra of the captured electrons in organic solids being observed in the interval of wave lengths of 500 to 2000 nm. With increasing polarity of the medium, λ_{max} is shifted toward the short-wave part of the spectrum. The sample heating from 4 to 77 K leads to the analogous effect. The same changes in the spectra of trapped electrons happen with increasing time of observation after the moment of their generation. A radiation yield of captured electrons is higher for the polar media as compared with the nonpolar ones and changes within 10^{-1} to 3 electrons per 100 cV. Also, it is necessary to note that a radiation yield e_t^- in crystals is sufficiently less than in glasses.

The lifetime of trapped electrons depends on a number of variables (temperature, presence of impurities, etc.) and can noticeably vary, sometimes reaching large quantities (hundreds of hours). The basic mechanisms of the generation of trapped electrons may be: localization of a polaron and capture of a conduction electron by a defect (e.g., by vacancy) in crystals. In addition, the electron capture occurs as a result of the Anderson localization (localization in a random field generated by structural defects) in disordered solids. If a trap is sufficiently deep, then in such potential wells discrete levels arise corresponding to the localized states of the electron. The ionization potentials of these levels, being arbitrary quantities, vary over a wide range, which explains the large width of the absorption band of trapped electrons.

As a rule, with increasing time of observation the absorption maximum band-wave length of trapped electrons is shifted toward a short-wave region. To explain this phenomena, two main hypotheses were put forward: "digging" of traps and "searching" of traps. In the first case, the trapped electron and polaron deepen the trap, polarizing the medium.[63,64] In the second case, it is assumed that a capture of the electron takes place in two stages: first, the capture is realized by shallow traps, and only then does the electron travel into more deep traps.[65] The information obtained from the ESR spectra permits the construction of an adequate model of a trapped electron. A description of these models can be found in Reference 66.

There is much less information on holes. The mobility of holes in solid organics is small. For alkanes[67] it is on the order of 10^{-2} $cm^2/V\ sec^{-1}$. The holes are rapidly localized generating cations. The localized holes (cations) absorb light in the visible and close infrared regions. So, for example, in 3-methylpentane, l_{max} equals 630 mm, and in squalane, 1400 nm at 72 K.[68] The localized holes in alkanes are unstable, as a result of which the ESR spectra were observed only in strongly polar matrices where the effective separation of charges takes place.[69]

2.2.3 The Excited States

Under the effect of ionizing radiation on a solid, excitation states also appear. These excitation states may be generated at direct excitation of molecules by ionizing radiation and at the recombination of electrons and ions.

The collision cross-section leading to the direct generation of excitation states is described by the Bete formula.[8] The excitation cross-section is proportional to the force of the oscillator of quantum transition, therefore, the singlet states are mainly excited, as in this case under transition there are no quantity changes of the total electron spin. Accounting for an electron exchange (as shown in References 70 and 71), the excitation cross-sections of triplet states rapidly decrease with the increasing energy of the electron E. It was found that it is proportional to E^{-3}. The calculations of the excitation cross-sections of triplet states are given in References 71 and 72. Since electrons with energy less than 100 eV prevail in the spectrum of the secondary electrons, the direct excitation of triplet states is rather effective.

Triplet states arise also as a result of the radiationless transition from singlet states (intersystem crossing). The effectiveness of the intersystem crossing is determined by a spin-orbital interaction and weakly depends on the aggregate state. This value lies in the limits of 0.1 to 1.[73]

The excited states may arise also in the recombination process of electrons and holes. Under the direct interband recombination of electrons and holes, it is necessary to consider the laws of conservation of energy and quasi-momentum. In those cases where the energy released under recombination may be partially removed (internal crossing, transfer of energy to the lattice), the exciton levels are populated (triplet and singlet states). The radiant recombination, as a rule, is not observed because usually the time constant of the internal crossing is sufficiently less than the radiant time of the annihilation of the highly excited state. In this case, we talk about a recombination through the exciton levels, the population of which is characteristic of solids.[74]

Under recombination, with the participation of recombination levels in connection with impurities and structure defects, the energy released may be insufficient for populating the triplet or singlet exciton levels. This follows from the fact that its value corresponds to the depth of the defect or impurity, which is usually less than the energy of the electron excitation of a molecule (the magnitude of the released energy, for example, is less than the value of the energy gap by the energy of electron affinity to the defect on which the electron is localized). The probability for recombination with the participation of impurity and structure defect levels is sufficiently higher than the probability of the direct interband recombination.

At low times of recombination, the spin correlation of the electron and cation (electron and hole) mainly remains. Therefore, in the given case, the singlet states are mainly generated. If a spin relaxation manages to occur before the moment of recombination, the probability for generating a triplet state is three times higher than a singlet one. Depending on the times of recombination and spin relaxation, the probability ratio for generating triplet and singlet states may vary from 0 to 3.[75,76]

Radiation yields of excitation states in organic substances are usually determined as a result of studying the radioluminescence of a substance, the influence of additions (acceptors of excitation and charge), magnetic field, etc. The best studied are electron excitations in aromatic hydrocarbons. For benzene, for example, the radiation yield of excitation states is of the order of ~6.5, with the yield of singlet states being ~1.5 and triplet states ~4.0.[58] The principal source of excitation states is a recombination.[77] The ratio of excitation states of benzene generated under direct excitation and as a result of recombination has been estimated as 2:3.4.[78] The radiant lifetime of excitation states is usually 10^{-8} to 10^{-9} sec for singlet states and 10^{-3} to 10^{3} sec for triplet states.[73] The other classes of organic substances have been studied much less. More detailed information on the radiation yields of excitation states in solid organic substances can be found in References 79 and 80.

2.3 SPATIAL NONUNIFORMITY OF ABSORBED ENERGY DISTRIBUTION

Under the effect of ionizing radiation, a transfer of energy to a substance proceeds by separate portions (discretely). The generated elementary excitations (electrons, holes, excitons, phonons, etc.) are localized in the vicinity of the primary electron trajectory and constitute its track. A classification of the track's energy structure was introduced in References 81 and 82. The space domain where 6 to 100 eV are absorbed is called a spur. A spur radius depends on the number of ion pairs in it and usually reaches several nanometers.[83] The average number of ion pairs in a spur equals approximately two, and the average energy released in a spur is ~30 eV.[84]

Under knock-out collisions, secondary electrons with energy from several hundred electronvolts to half of the incident electron energy appear. Such secondary electrons create their own tracks branching from the primary electron track. If the energy absorbed by the medium in such a track is 100 to 500 eV, it is called a blob. Blobs are characterized by a larger content of ion pairs and dimension as compared to spurs. Short tracks represent wider tube domains where 500 to 5000 eV of the secondary electron energy have been absorbed. Blobs differ from short tracks in that under generation of blobs the energy of a secondary electron is insufficient to overcome the attraction of its ion. Under the energy of a secondary electron larger than 5000 eV, a branch track appears whose structure is similar to the main track.

A distribution of energy in spurs, blobs, and short tracks depending on the electron energy was calculated in Reference 82. If the electron energy is 1 MeV, 65% of the energy lost goes toward generating spurs, 15% for generating blobs, and 20% for generating short tracks. With increasing electron energy, losses from generating spurs increase. The length of a free path of a high-energy electron (distance between acts of an energy transfer) greatly exceeds the dimensions of the domain of an absorbed energy localization. So, for example, for an electron with the energy 1 MeV,[85] the interspur distance is 500 nm and the dimension of the domain of an absorbed energy localization is of the order of 5 to 10 nm. Methods for calculating the degradation spectra and energy structure of fast electron spectra are considered in References 86 and 87.

Absorption of the energy of high-energy radiation by the medium is determined by the energy transferred to a substance in a unit of a track length. This quantity is called a linear energy transfer (LET). In the case of a large linear energy transfer, which is characteristic of heavily charged particles, the interspur distance decreases and they merge.

The track of a heavily charged particle represents a tube domain consisting of the merged spurs generated by a primary particle (the "core") and shell (penumbra) consisting of the secondary electrons' tracks.[88] The loss of energy per unit core volume is much higher than in the penumbra. A core dimension depending on a particle energy is 0.1 to 10 nm and that of the penumbra 10 to 105 nm. The peculiar features of heavily charged particle tracks are used for producing nuclear filters (see Chapter 9).

When considering the degree of absorbed energy localization, it is necessary to take into account its delocalization owing to the quantum character of elementary excitations of a solid and high-energy particle interacting with it. Indeed, as shown in Chapter 1, electrons and holes in solids (especially in crystals) are greatly delocalized, i.e., made collective by all atoms of the lattice. The same can be said about the other elementary excitations of a solid — polarons and excitons. They are also "spread" over the crystal.

The possibility and place of localization of such excitations depend on transfer processes and defectiveness of a solid lattice. The localization domain may be apart at a large distance from the place of the energy absorption. A migration length of elementary excitations is 10 to 100 nm (see Section 2.1). For example, such collective excitation states as plasmons are localized in the domain with linear dimensions 30 nm.[89]

When a particle interacts with a substance, there are limits to the precision of determining a particle position in space and the magnitude of its momentum (energy). These limits are imposed by means of a principle of uncertainty. If in the process of the energy-to-substance transfer a component of the particle momentum is determined with precision to the value of the transferred momentum, then according to a principle of uncertainty the position of a particle is determined with the precision Δx. This ratio can be written in the form which is usually utilized in radiation physical chemistry:

$$\Delta x \geq \hbar \, v_x / 2\Delta E,$$

where v_x is the particle velocity, ΔE is the energy transferred to a substance.

It is easy to see that the smaller the value of the transferred energy, the higher the uncertainty in determining the place of its absorption. Table 2.1 gives the data illustrating the uncertainty for the transferred energy value $\Delta E = 15$ eV.

Table 2.1 Value of Delocalization Absorption Energy as a Function Energy of High-Energy Particle

	E, eV						
	10^1	10^2	10^3	10^4	10^5	10^6	10^7
$v \cdot 10^6$, m/sec	1.87	5.93	18.7	58.5	164	282	300
Δx, nm	0.08	0.26	0.82	2.56	7.2	12.3	13.1

It can easily be seen that for the energies 1 MeV, a spatial uncertainty of absorption is much higher and makes up ~12 nm. For low-energy electrons with energy of 10 eV, this value is negligible. During absorption of the electron energy $\Delta E = 7.5$ eV by the medium, the delocalization is twice as much as compared to the values given in Table 2.1. In the case of absorbing the energy $\Delta E > 100$ eV, which is characteristic of knock-out collisions, the losses of energy are of a virtually localized character. Thus, the delocalization of energy by an absorbed medium can be sufficient to lead to erosion of the tracks and, correspondingly, the disappearance of track effects.

2.4 ABSORPTION OF ENERGY IN HETEROGENEOUS SYSTEMS

Heterogeneous systems are characterized by the presence of a phase interface. This interface is the surface that causes the majority of regularities of heterogeneous processes. The absorption of energy in such systems has a number of peculiar features related to the distortion of electron equilibrium at the phase interface by means of various effects. Besides a capture of charges and energy by surface levels and defects, formation of a solid spatial charge as well as a transfer of energy and charge over the surface play an important part.

A distortion of electron equilibrium near the surface of the phase interface under irradiation of heterogeneous systems first takes place due to the fact that the secondary electrons generated in one of the media can find themselves in the other. Depending on the properties of a system, the interphase exchange of secondary electrons leads to both an increase in energy absorption by a unit mass in the near-surface layer as compared to the volume and a decrease in absorption. The phases may differ greatly in braking capability, which is defined by effective atomic numbers of adjoining substances. So, for example, a braking capability will vary noticeably at the phase interface in such systems as solid-gas or organic solid-metal. As a result, the secondary electrons generated in a medium with a large effective atomic number (phase 1) will not be compensated by the secondary electrons incoming from a medium with a smaller effective atomic number (phase 2). In this case, in the near-surface layer of phase 1, the unit mass will absorb a smaller amount of energy than in the volume, while in the near-surface layer of phase 2, a larger amount.

In Reference 90 it was shown that in exposing a polymer film to bremsstrahling and γ-radiation under the contact of the film with vacuum or gaseous phase, the energy value absorbed by a unit mass depends on the film thickness and radiation energy: the thinner the film, the less the absorbed energy. For radiation having the energy 1.25 MeV (Co^{60}), the energy absorbed in the film 200 μm thick makes up ~30% of the solid one. With increasing film thickness, the absorbed energy reaches the value characteristic of the volume. With increasing quantum radiation energy, the thickness of a layer in which a nonuniformity of energy absorption is observed increases. This follows from the fact that the length of a free path to secondary electrons increases, a part of which may in this case leave the sample.

In the near-surface layer of the medium with a less effective atomic number (phase 2), an additional absorption of radiation energy as compared to the volume is explained in various ways. Here it is not only the secondary electrons that transferred from the phase with a larger effective atomic number (phase 1) that are important. Also important in the given case are the processes of inverse scattering in phase 1 of the secondary electrons which are generated in phase 2 (see Figure 2.1). The probability for the secondary electrons to be scattered increases with decreasing energy and an increasingly effective atomic number of the medium.

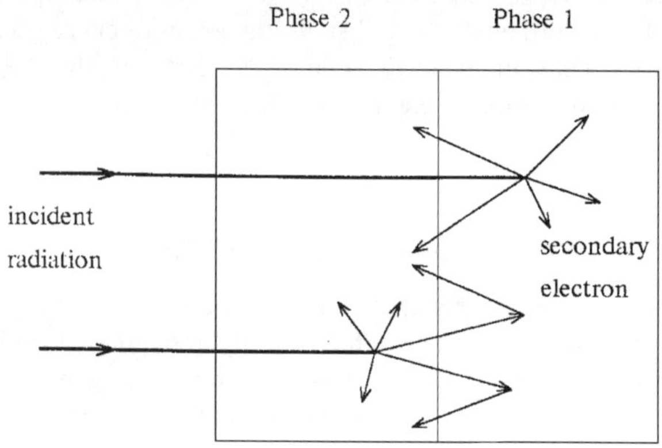

Figure 2.1 Sheme of high-energy absorbence in heterogeneous systems.

In Reference 91, calculations were made of the absolute yields of electrons scattered inversely to the direction of the incident radiation from the target whose width provides a total energy absorption. The yield of electrons calculated for one absorbed quantum (1.25 MeV) is 0.25 for aluminium and 0.4 for lead. The absorbed near-surface energy may exceed several times the absorbed energy in homogeneous conditions. Such results for heterogeneous systems were observed repeatedly.[92,93]

A redistribution of absorbed energy in heterogeneous systems also takes place as a result of an energy and charge transfer over the phase interface (see Chapter 3). The elementary excitations of a solid may migrate at sufficient distances (up to hundreds of nanometers). In the scientific literature, various mechanisms of energy transfer over the surface were treated: resonance exchange of excitation-energy between phases,[94,95] capture and decay of excitons at the medium interface, transfer of electrons and holes over the boundary and their further recombination,[97,98] and transfer of atoms and radicals over the surface.[99]

The presence of a spatial charge domain at the phase interphase noticeably affects electron processes in the near-surface layer. So, for example, the velocity of the surface recombination of a charge carrier changes greatly depending on the value and sign of the surface potential Y_s (see Chapter 1). This leads to a ratio of electron excitation and ionization states different from the volume.

Thus, in heterogeneous systems the distribution and spectrum of elementary chemically active excitations generated under energy absorption differ much from the homogeneous medium. This leads to the conclusion that the yield of radiation-chemical transformations in heterogeneous systems in many cases is higher as compared to the homogeneous medium. The mechanisms of radiation-chemical conversions in heterogeneous and homogeneous systems also may differ significantly. Some of the above-mentioned peculiar features of heterogeneous radiation-chemical processes will be treated in Chapter 8.

REFERENCES

1. Hart, E.J., Platzman, R.L., in *Mechanisms in Radiobiology,* Errera, M., Forssberg, A., Eds., Academic Press, New York, 1, 17, 1961.
2. Evans, R.D., *Radiation Dozimetry,* 2nd ed., Vol. 1, Attix, F.H., Roesch, V.F., Eds., Academic Press, New York, 1968, 93.
3. Landau, L.D., Lifshitz, E.M., *Quantum Mechanics,* Pergamon Press, Elmsford, NY, 1981.
4. Evans, R.D., *In The Atomic Nucleus,* Schif, L.I., Ed., McGraw-Hill, NY, 1955.
5. Jackson, J.D., *Classical Electrodynamics,* 2nd ed., John Wiley & Sons, New York, 1975.
6. Massey, H.S.W., Burhop, E.M.S., *Electronic and Ionic Impact Phenomena,* Clarendon Press, Oxford, 1952.
7. Mott, Massey, H.S.W., *The Theory of Atomic Collision,* Clarendon Press, Oxford, 1965.
8. Bethe, H., Ashkin, J., *In Experimental Nuclear Physics,* Vol. 1, Segre, E., Ed., John Wiley & Sons, New York, 1953, 166–357.
9. Anlen, S.P., Theoretical and experimental aspects on the energy loss of relativistic heavily ionizing particles, *Rev. Mod. Phys.,* 52, 121, 1980.
10. Walske, M.C., Stopping power of L electrons, *Phys. Rev.,* 101, 940, 1956.
11. Bichsel, H., Straggling of heavy charged particles: comparison of born hydrogenic-wave-function approximations with free-electron approximation, *Phys. Rev.,* 1, 2854, 1970.
12. Fano, U., Deep penetration of radiation, *Am. Rev. Nucl. Sci.,* 13, 1, 1963.
13. *Atomic Inner-Shell Processes,* Graserman, B.N., Ed., Academic Press, New York, 1975, 1.
14. Matveyev, V.I., Parilis, E.Ñ., Shake-off process at the electron transition in atoms, *Uspekhi Fizicheskikh Nauk,* 138, 573, 1982.
15. Karlson, T.A., Krause, M.O., Electron shake-off resulting from k-shell ionization in neon measures function of photoelectron velocity, *Phys. Rev. A.,* 140, 1057, 1965.
16. Fano, U., Normal modes of a lattice of oscillations with many resonances and dipolar coupling, *Phys. Rev.,* 118, 451, 1960.
17. Williams, H.W., Hanom, R.N., Arukawa, E.T., Painter, L.N., Birkhofl, R.D., Excitation of plasmons and interband transitions by electrons, *Int. J. Radiat. Phys. Chem.,* 1, 95, 1975.
18. Kaplan, I.G., Modern trends in the development of theoretical radiation chemistry, *High Energy Chem.,* 17, 210, 1983.
19. Kutcher, C.J., Green, E.S., A model for energy deposition in liquid water, *Radiat. Res.,* 67, 408, 1976.
20. Kaplan, I.G., Miterev, A.M., Specific of ionised radiation interaction with molecular media and role of the tracks in the radiation chemistry, *Russ. Chem. Rev.,* 55 N 5, 713–742, 1986.
21. Platzman, R.L., Superexcited state of molecules, *Radiat. Res.,* 17, 419, 1962.
22. De Heer, F.G., Superexcited of molecules produced by electron impact, *Intl. J. Radiat. Chem.,* 7, 137, 1975.

23. Hatano, Y., Excitation from superexcitation of hydrocarbons, estimation of their yields by the optical approximation, *Bull. Chem. Soc.,* 41, 1126, 1968.
24. Makarov, V.I., Polak, L.S., Excited and superexcited electronic states of atoms and molecules and role they play in radiation chemistry, *High Energy Chem.,* 4, 3, 1970.
25. Nelson, R.S., Tompson, M.W., Collision sequences in the crystals of copper, silver and gold revealed by sputlering in energetic ion beams, *Proc. Roy. Soc.,* A 259, 458, 1961.
26. Silsbe, R.H., Focusing in collision problems in solids, *J. Appl. Phys.,* 28, 1246, 1957.
27. Leibfried, G., Correlated collisions in a displacement spike, *J. Appl. Phys.,* 30, 1388, 1959.
28. Lehmann, C., Leibfried G., Fokussirence 110-stobfolgen in Flächenzentrierten Kristallen bei Kleinen Winkeln, *Z. Phys.,* 162, 203, 1961.
29. Gemmel, D.S., Channelling and related effects in the motion of charged particles through crystals, *Rev. Mod. Phys.,* 46(1), 129–227, 1974.
30. Varley, J.H.O., A mechanism for the displacement of ions in an ionic lattice, *Nature,* 174, 886, 1954.
31. Markham, S.J., Speculation on the formation of F-centers during irradiation, *Phys. Rev.,* 88, 500, 1952.
32. Lehman, Chr., *Interaction of Radiation with Solids and Elementary Defect Production,* North-Holland, Amsterdam, 1977.
33. Stoneham, A.M., *Theory of Defects in Solids,* Clarendon Press, Oxford, 1975.
34. Inokuti, M., Inelastic collisions of fast charged particles with atoms and molecules — the Bothe theory revisited, *Rev. Mod. Phys.,* 43, 297, 1971.
35. Inokuti, M., Itikawa, Y., Turner, J.E., Adenda: in inelastic collisions of fast charged particles with atoms and molecules, the Bothe theory revisited, *Rev. Mod. Phys.,* 50, 23, 1978.
36. Tan, K.N., Brion, C.E., van der Leeuw Ph.E., van der Wiel, M.J., Absolute oscillator strengths (10-60 eV) for the photoabsorption, photoionization, and fragmentation of H_2O, *J. Chem. Phys.,* 29, 299, 1978.
37. Grosse, H.J., Bothe, H.H., Ionisierungs querschnitte von Kohlen wasserstoffen und deren sauerstoffderivaten, *Z Naturforch.,* B23a, 1583, 1968.
38. Grosse, H.J., Bothe, H.H., Ionisierungsquerschnitte von stickstoff- und halogenkohlenwasserstoffderivaten, *Z Naturforch.,* B25a, 1970.
39. Myers, I.T., Ionization, in *Radiation Dosimetry,* 2nd ed., Attix, F., Rolsch, W.C., Eds., Academic Press, New York, 1968, 317–330.
40. Huang, S.S.-S., Freeman, G.R., Effect of density on the total ionization yields in X-irradiated argon, krypton and xenon, *Can. J. Chem.,* 55, 1838, 1977.
41. Brattacharya, D., Willard, J.E., Spectra, yields, photolability of cations, carbanions, radicals and scavenger anions in g-irradiated alkanes glasses, *J. Phys. Chem.,* 84, 146, 1980.
42. Noziereas, P., Pines, D., Electron interaction in solid. Corrective approach to the dielectric constant, *Phys. Rev.,* 109, 762, 1958.
43. Platzman, R.L., Subexcitation electrons, *Radiat. Res.,* 2, 1, 1955.
44. Samuel, A.H., Magee, J.L., Theory of radiation chemistry. II. Trackeffect in radiolis water, *J. Chem. Phys.,* 21, 1050, 1953.
45. Warman, J.M., in *The Study of Fast Processes and Transient Species by Electron Pulse Radiolysis,* Baxendale, J.H., Busi, F., Eds., Reidel, Dordrecht, 1982, 86, 437–527.
46. Onzager, L., Initial recombination of ions, *Phys. Rev.,* 54, 554, 1938.
47. Mozumder, A., Magee, J.L., Theory of radiation chemistry. VIII. Ionization of nonpolar liquids by radiation in the absence of external electric field, *J. Chem. Phys.,* 47, 939, 1967.

48. Freeman, G.R., The radiolysis of aliphatic and alicyclic hydrocarbons, *Radiat. Res. Rev.*, 1, 1, 1968.
49. Vavilov, V.S., *Radiation Influence on the Semiconductors*, Fizmatgiz, Moscow, 1963 (in Russian).
50. Van Roosbroeck, W., The transport of added current carriers in a homogeneous semiconductor, *Phys. Rev.*, 91(2), 282, 1953.
51. Fugol', I. Ya., Excitons in rare-gas crystals, *Adv. Phys.*, 27(21), 201, 1978.
52. Warman, J.M., Kunst, M., De Haas, M.P., Electron and proton conduction in ice, *J. Electrostatics*, 12, 115, 1982.
53. Huang, T., Eisele, I., Kevan, L., Electron driff Hall mobility in g-radiated 10M NaOH glassy alkane ice, *J. Chem. Phys.*, 59, 6334, 1973.
54. Schmidt, K.H., Buck, W.L., Mobility of the hydrated electron, *Science*, 151, 70, 1966.
55. Gutmann, F., Lyons, L.F., *Organic Semiconductors*, John Wiley & Sons, New York, 1967.
56. Schmidt, W.F., Allen, A.O., Electron mobility in solid neopentane, *J. Chem. Phys.*, 50, 2667, 1972.
57. Namba, H., Shinsara, K., Hatano, Y., Effect of n-butane impurity on ion mobility and electron recombination rate constant in solid neopentane, *J. Chem. Phys.*, 70, 5331, 1979.
58. *Radiation Chemistry of Hydrocarbons*, Foldiak, G., Ed., Elsevier, Amsterdam, 1981.
59. Namba, H., Chiba, M., Nakamura, Y., Tezuka, T., Shinsaks, K., Hatano, Y., Electron mobilities in liquid and solid cyclohexane, *J. Electrostatic*, 12, 79, 1982.
60. Naruyama, Y., Funabashi, K., Electron drift mobility in the solid 3-methylpentane, *J. Chem. Phys.*, 56, 2342, 1972.
61. Ershov, B.G., Pikaev, A.K., Stabilization of electrons in low-temperature radiolysis polar systems, *Adv. Chem. Sep.*, 81, 1, 1968.
62. Kevan, L., in *Advances in Radiation Chemistry*, Burton, M., Magee, J.L., Eds., Wiley Interscience, New York, 1974, 181.
63. Ogasawara, M., Kevan, L., Laser photoionization study of time dependent spectral shifts of localized electrons in ethanol glass. Temperature effects, *J. Phys. Chem.*, 82, 378, 1978.
64. Dolivo, G., Kevan, L., Optical absorption spectra of localized electrons generated at 1.6 K in polar matrices: evidence for presolvated electrons, *J. Chem. Phys.*, 70, 2599, 1979.
65. Baxendale, J.H., Sharpe, P.H.G., Electron solvation in alcohols at 77 K after pulse radiolysis, *Chem. Phys. Lett.*, 39, 401, 1976.
66. Funabashi, K., in *Advances in Radiation Chemistry*, Wiley Interscience, New York, 1974, 4, 103.
67. Hermann, R., Brede, O., Boss, J., Mehucrt, R., The mechanism of solute singlet state formation in irradiated liquid alkanes, *Ber Bunsenges. Phys. Chem.*, 84, 814, 1980.
68. Theather, G.G., Klassen, N.V., Positive ions in alkane and mixed alkane glasses, *J. Phys. Chem.*, 85, 3044, 1981.
69. Toriyama, K., Numone, K., Iwasako, M., Structures and reactions of radical cations of some prototype alkanes in low temperature solid stadied ESR spectroscopy, *Chem. Phys.*, 77, 12, 1982.
70. Ochkur, V.I., On the Born-Oppenheimer method in atomic collision theory, *Sov. Phys. JETF*, 45, 734, 1963.
71. Rudge, M.R., The calculation of exchange scattering amplitudes, *Proc. Phys. Soc.*, 85, 607, 1965.
72. Matsuzawa, M.J., Electron-impact excitation cross sections of the lowest-lying triplet states of benzene, *Chem. Phys.*, 51, 4705, 1969.
73. Calvert, J.G., Pitts, J.N., *Photochemistry*, John Wiley & Sons, New York, 1966.
74. Frankevich, E.L., Ionic and electronic processes in condensed hydrocarbons under radiation, *Russ. Chem. Rev.*, 35, 1161, 1966.

75. Magee, J.L., Huang, J.T., Triplet of formation in recombination in spurs, *J. Phys. Chem.,* 76, 3801, 1972.
76. Magee, J.L., Huang, J.T., Comment on "triplet of formation in recombination in spurs", *J. Phys. Chem.,* 78, 310, 1974.
77. Merkel, P.B., Hamill, W.H., Luminescence from then solid films at 77 K under low-energy electron bombardment, *J. Chem. Phys.,* 54, 1695, 1971.
78. Hentz, R.R., Shermann, W.V., Charge scavenging and energy transfer in a g-radiolysis of benzene solution, *J. Phys. Chem.,* 73, 2676, 1969.
79. Àlfimov, M.W., The triplet states in radiation chemistry, *High Energy Chem.,* 6, 3, 1972.
80. *The Formation and Role of Excited States in Radiolysis,* special issue of *Int. J. Radiat. Phys. Chem.,* 8, 1, 1976.
81. Mozumder, A., Magee, J. L., Model of tracks of ionizing radiations for radical reaction mechanisms, *Radiat. Res.,* 28, 203, 1966.
82. Mozumder, A., Magee, J. L., Theory of radiation chemistry. VII. Structure and reaction in low LET tracks, *J. Chem. Phys.,* 45, 3332, 1966.
83. Copeland, D.A., Kester, N.R., Jortnes, J., Excess electrons in polar solvents, *J. Chem. Phys.,* 53, 189, 1970.
84. Magee, J.L., Chatterjee, A., A spur unfolding model for the radiolysis of water, *Radiat. Phys. Chem.,* 15, 125, 1980.
85. Dewhurst, H.A., Samuel, A.H., Magee, J.L., A theoretical survey of the radiation chemistry of water and aqueous material, *Rad. Res.,* 1, 62, 1954.
86. Mozumder, A., in *Advances in Radiation Chemistry,* Button, M., Magee, J., Eds., Wiley Interscience, New York, 1969, 1.
87. Kaplan, I.G., Skurat, W.A., Energy distribution of second electrons and track effects, *High Energy Chem.,* 6, 252, 1972.
88. Magee, J.L., Chatterjee, A., Radiation chemistry of heavy-particle tracks, I. General considerations, *J. Phys. Chem.,* 84, 3529, 1980.
89. Kaplan, I.G., Miterev, A.M., On the problem of ionizing radiation energy delocalization in a molecular medium, *Dokl. Phys. Chem.,* 280, 127, 1985.
90. Johns, H.E., Cunningham, J.R., *The Physics of Radiology,* 4th ed., Charles C Thomas, Springfield, IL, 1983.
91. Hine, J., Scattering of secondary electrons produced g-rays in materials of various atomic numbers, *Phys. Rev.,* 82, 755, 1951.
92. *Proceedings of Third Symposium on Microdosimetry,* Ebert, H.G., Ed., Euroatom, Luxemburg, 1972.
93. *Proceedings of Fourth Symposium on Microdosimetry,* Booz, J., Ebert, H.G., Waker, A., Ed., Euroatom, Luxemburg, 1974.
94. Myasnikov, I.A., Grigoriev, E.I., Tsivenko, V.I., Electro-excited atoms and molecules in gas-solid state systems, *Russ. Chem. Rev.,* 55, 161, 1986.
95. Akimov, A.Z., Cherkasov, Y.A., Cherkashin, M.I., *Sensibilized Photoeffect,* Nauka, Moscow, 1980 (in Russian).
96. Agranovich, V.M., Galanin, M.D., *Energy Transfer of Electron Excitation in Condensed Matter,* Nauka, Moscow, 1978 (in Russian).
97. Sagert, N.H., Reid, J.A. Robinson, R.L., Radiolysis in the absorbed state. III. Methyl iodide absorbed on silica gel, *Can. J. Chem.,* 48, 17, 1970.
98. Zhabrova, G.M., Vladimiriva, V.I., The main regularity and role of electronic factors in catalytic processes which follow under action of ionizing radiation, *Russ. Chem. Rev.,* 38(4), 711–739, 1969.
99. Dyne, P., Sagert, N.H., Exchange reactions in the radiolysis of absorbed hydrocarbons, *Nature,* 210, 1153, 1966.

3 PRIMARY CHEMICAL PROCESSES IN THE IRRADIATED SOLID

As the substance is irradiated, the primary processes of formation of elementary excitations take place, and chemically active particles are accumulated. The latter include ions, electron-excitation molecules and atoms, radicals, and stabilized electrons. All these particles readily enter into the chemical reactions considered in this chapter.

It should be noted that these issues are treated in the following four chapters, also; however, those chapters emphasize how a medium influences the course of radical reactions. Here, the consideration is of a more general character.

3.1 ION REACTIONS

Molecular cations formed under irradiation can undergo ion-molecular reactions with neighbouring molecules. These reactions lead to the formation of radicals which are stabilized in the solid phase at low temperatures and registered by the ESR method.

At the same time a sufficient fraction of cations formed in vibrational- and electron-excited states. Dissociation of excited ions generated radicals. A radical yield usually serves as a quantitative criterion of a substance's chemical resistance to radiation effect.

Until now it was thought that decay of excited ions in the solid phase is unlikely and formation of radicals is determined by the contribution of ion-molecular reactions. Two circumstances served as a prerequisite for this — a large dissipation rate of excitation energy in the condensed phase and large cross-section values of ion-molecular reactions in the gaseous phase. A zero proximity of activation energies of ion-molecular reactions in the gaseous phase would seem to lead to an increasing role of these reactions at low temperatures. However, using the results of measurements in the gaseous phase to analyze processes occurring in the solid phase is not exactly valid.

First, under charge transfer in condensed substances, the medium is rearranged (repolarized) depending on energy that is used (see Section 5.4). Secondly, the potential surface which determines the energy of reaction activation is determined in gases only by interaction of reagents. For solids, the potential surface must be formed with regard to molecular environment. So, for example, in Reference 1 a potential surface for reaction of ion polymerization is constructed with regard to molecular environment.

The activation energy of ion-molecular reactions in the solid phase is determined not only by interaction of the ion with the molecule which participates in the reaction but also by the energy of the medium rearrangement and matrix stiffness. Hence, this activation energy will not necessarily be small.

On the other hand, the probability of the monomolecular decay of an excited ion is dependent on how the processes of energy dissipation and dissociation compete with each other. References 2 and 3 have analyzed the dependence of the dissociation rate constant of excited molecular systems on the energy of bond rupture. It has been shown that at the bond energies ≤2 eV, the dissociation process can compete successfully with energy dissipation. The values of binding energies in cation-radicals allow the conclusion to be made that in some cases the decay of excited ions is essential and sometimes also a determining mechanism of the free radicals formation in the solid phase.

Experimental results demonstrating this conclusion are available. Reference 4 discusses a study of radiolysis and photolysis of methylene chloride and bromide in an argon matrix at 15 K. Parent and product ions and radicals were observed. By exciting parent ions by light with different wavelengths, the same product ions and radicals were produced as in radiolysis. This shows that the process of monomolecular decay of excited ions proceeds effectively enough in the solid phase. From the coincidence of products in radiolysis and photolysis, a conclusion about the process mechanism can be made. Both in photolysis and radiolysis the dissociation of excited particles is the primary mechanism. It should be pointed out, however, that under conditions of matrix isolation relaxation is difficult; therefore, the dissociation effectiveness of excited cations in the matrix proper may be less.

A more definite answer to the question about the importance of monomolecular decay of cations in the solid phase is given in Reference 5, which describes an investigation into the radiolysis mechanism of butyl chloride under conditions of matrix isolation and in the matrix proper. As a result of irradiation, both cases have similar absorption spectra which belong to the parent and product cations. Addition of diphenyl decreases the ion concentration by a similar ratio, the cation $C_{12}H_{10}^{+\cdot}$ being formed. This means that the parent and product ions are produced from one predecessor, the excited parent ion. In addition, the pike intensities in product spectra of buthylchloride and buthylbromide are close to the intensities of their mass-spectra. These spectra, as is known, are produced in the absence of collisions when a monomolecular decay is the only possible mechanism of reaction.

At the same time, the monomolecular decay of excited alkane ions (in spite of a small energy bond value) apparently does not occur. Stabilization of a number of alkane cation-radicals was observed in the matrices of SF_6, $CFCl_2CF_2Cl$ as a result of γ-radiation at a temperature of 4.2 K.[6] It is interesting to note that under matrix heating up to 100 K, a conversion was observed of cation-radicals $C_2H_6^{+\cdot}$, $C_3H_8^{+\cdot}$, $(CH_3)_3CH^+$, $(CH_3)_4C^+$ into radicals C_2H_5, C_3H_7, $CH_2CH_2(CH_3)_2$, $CH_2C(CH_3)_3$, with anions SF_6^- and $CFCl_2CF_2Cl^-$ diminishing simultaneously. Thus, radicals are formed as a result of recombination of cation-radicals with anion-radicals of the matrix under the origin of mobility of a captured electron. In the case of alkanes, among the two processes of energy dissipation and dissociation of excited cation-radicals, the process of dissipation probably prevails.

3.2 REACTIONS OF ELECTRON-EXCITED MOLECULES

Dissociation of electron-excited molecules is an important radiation-chemical process bringing about radicals formation. It is known (see Chapter 2) that the radiation yield of excited molecules is approximately equal to the yield of ionized molecules.[7,8] When studying solid-phase radiation-chemical reactions, one should take into account that the excitation energy transfer is a characteristic phenomenon that accompanies a chemical process. The intermolecular and intramolecular transfer of energy is the intermediate stage between radiation energy dissipation processes and chemical reactions. The influence which this stage has on conversion regularities under a radiation effect can be decisive. This section will consider chemical effects of exciton and intramolecular transfer of energy.

3.2.1 Radiation-Chemical Reactions of Organic Molecules in an Inert Matrix

A convenient technique for separating reactions of electron-excited molecules is as follows. A sample is prepared in which organic molecules are diluted with an essential amount of inert gas. After freezing, an inert gas matrix containing organic molecules (methane, propane) is formed. When a xenon matrix at temperature 77 K was used, the organic content was from 1 to 10 mol%.[9,10]

The xenon matrix surface was bombarded by atoms of xenon and argon excited to a metastable state. The energy of the excited atom of xenon is 8.44 eV and that of argon, 11.62 eV. The energy of the lower exciton level in xenon cryocrystal is 8.4 eV.[11] It is easy to see that this energy is not enough for ionizing a methane molecule (the ionization potential of methane is 12.7 eV), but quite enough for electron excitation of the triplet state. The energy of the excited singlet level of methane is of the order of 10 eV, and that of the triplet, 6 eV. Hence, under given conditions, excitons are propagated in xenon matrix.

As a result of bombardment in the matrix containing methane, methyl radicals are accumulated (Figure 3.1). For the matrix containing 1 mol% of methane, the estimation gives the width of the layer approximately equal to $4 \cdot 10^{-6}$ cm. The radius of exciton capture by methane molecule is 2.5 Å.

When the matrix contains propane, curve 3 corresponds to radicals accumulation (Figure 3.1). The thickness of the reaction layer and the capture radius for these tests are not managable for estimation. Propane radicals may form not only under dissociation of excited propane molecules but also as a result of H-atom reactions with propane molecules. We shall note that the analogous reaction for methane molecules ($CH_4 + H \rightarrow CH_3 + H_2$) is endothermic and therefore does not occur at low temperatures.

A formation of radicals under the conditions of matrix isolation (methane-xenon) was studied also under γ-radiation.[10] Experiments were carried out at a temperature of 77 K at a methane concentration in the range of 0.5 to 5 mol%. The dependence of radical concentration on time is shown in Figure 3.2. The radius of exciton capture by methane molecule and its lifetime are close to the values of the analogous parameters when bombarding the matrix surface by excited atoms of xenon. Besides that, a radiation yield of radicals which calculated per energy absorbed by methane only is in the range of 86 to 250 in dependence of methane

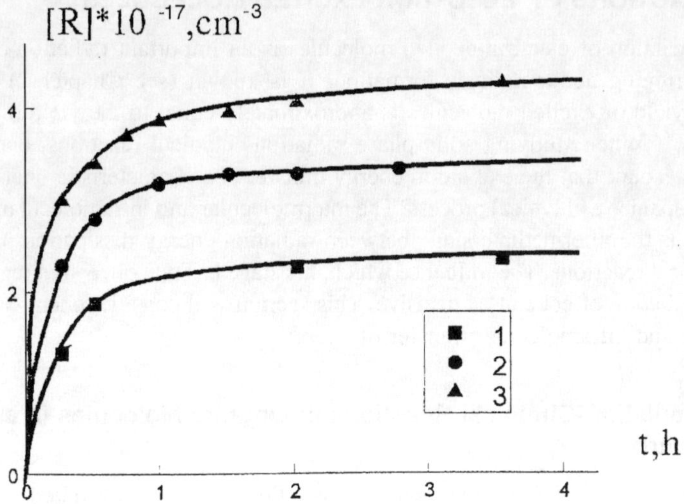

Figure 3.1 Amount of methyl radicals as a function of surface bombardment time containing methane and propane matrix by metastable xenon atoms. Molar fraction of propane is 10%(1) and methane is 1%(2), 10%(3).

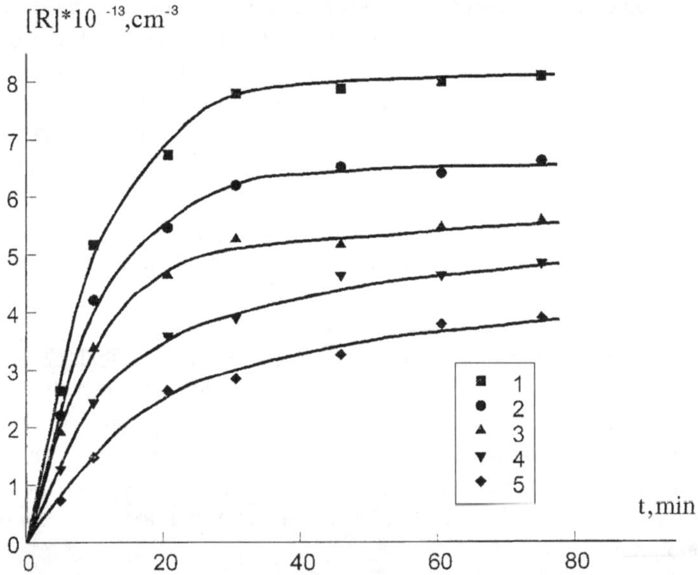

Figure 3.2 Amount of methyl radicals as a function of surface bombardment time containing methane matrix by metastable xenon atoms. Molar fraction of methane is 0, and benzene is 0.25%(1), 0.33(2), 0.5(3), 0.664(4), and 1.0(5).

concentration. It is clear that for 100 eV of absorbed energy, such a yield is impossible. This demonstrates the matrix participation in the radiation process. Thus, the exciton mechanism of radicals formation is proved. The process of dissociation of excited molecules is a consequence of the exciton effect.

Additional data on the radiolysis mechanism of the above-mentioned systems were obtained by incorporating aromatic additions[12] into them. In particular, a small addition of benzene markedly reduces the radiation yield of methyl radicals. Thus, the formation of radicals in a methane-content xenon matrix under bombardment by metastable xenon atoms convincingly demonstrates the dissociation of electron-excited molecules.

In general, there is a large number of examples of chemical conversions of electron-excited molecules which are sensitized by an inert gas matrix. Papers present various evidences of a noticeable influence of matrix energy transfer on chemical processes.

The band structure and dynamics of atomic excitation in the crystals of noble gases have been well studied.[11,12] In many cases, it was possible to identify an elementary chemically active, radiation-induced excitation and to abstract the primary processes, the migration and capture of excitons by impurities (organic molecules) and decay of these molecules to radicals.

When conventional or deuterated methane is exposed to γ-radiation in argon and krypton at 4.2 K, formation and stabilization of H atoms and CH_3 (CD_3) radicals occurred.[14,15] The radiation yield of radicals per energy absorbed only by methane would be from 500 to 1000 with a methane content in the matrix from 10^{-1} to 10^{-2} mol. These results convincingly demonstrate that hydrogen atoms and methyl radicals are formed as a result of γ-radiation energy absorption by the inert matrix and the following transfer of energy to impurity (methane).

The sensitization effect of methane decomposition in a xenon matrix at a temperature of 77 K was observed even at a high content of methane in the γ irradiated matrix.[16] Similar results were obtained under irradiation (10 K) of Xe, Kr, and Ar matrices containing 0.5 mol% CH_4, C_2H_6, 3MP-h14, 3MP-d14, i-C_4H_{10}, with X-ray[17] and at γ-radiolyse of xenon matrices containing 1 mol percent CH_4, C_2H_6, CH_3CD_3, C_3H_8, i-C_4H_{10}, C_2H_4.[18] These works show that the radiation yields of stabilized H-atoms and radicals are equal. It speaks in favor of the unimolecular decay mechanism of the substances studied.

A deviation from "the rule of additivity" (the absorption of energy by ionizing radiation is proportional to the electronic fragments of irradiated substances) is explained by the matrix positive-charge transfer ("hole") over the impurity.[15] It should be kept in mind that in the case of a xenon matrix this is impossible, since the ionization potential of xenon atom is equal to 12.1 eV and that of methane, 13 eV. At the same time, the transfer of a "hot hole" ($Xe^+(2P_{1/2})$ which corresponds to the energy 13.43 eV is not forbidden. However, a molecular ion CH_4^+ may be rapidly neutralized by electron transfer from Xe without bond rupture in a methane molecule.

It was also assumed that a hydrogen atom transfers from methane to the nearest ion of Xe^+ with further neutralization: $XeH^+ + e^- \rightarrow Xe + H$.[19] In this case, a molecular cation (XeH^+) must effectively capture an electron of the resultant radical. Such a reaction is exothermic.

We shall note that a mechanism of radicals formation under radiant energy transfer in these systems is impossible. The excitation of the impurity by luminescence of the free excitons in the matrix is refuted by data showing that the exciton luminescence in noble gases is totally suppressed in the presence of even a small amount of impurity.[11]

In Reference 20, the authors conducted a study of photolysis and radiolysis of 2-methylbutane in an argon matrix (at temperatures of 20 and 77 K) and one of krypton (at a temperature of 77 K). It was found that the products of photolysis and radiolysis are the same, and under radiolysis an energy transfer from the matrix to the 2-methylbutane impurity has been discovered.

A possible chemical effect of the excitons of noble gases was demonstrated when exciting excitons by both vacuum ultraviolet (synchrotron radiation)[19-21] and interaction of the matrix surface with metastable atoms of noble gases.[9,10] It was shown that the triplet excitons, formed in a xenon matrix (energy 8.4 eV), are capable of ionizing a benzene molecule's impurity (1 mol%) at a temperature of 40 K which is manifested in the photoconductivity of such a sample.[21-23] Parameters are determined, such as the exciton capture rate constant by a benzene molecule leading to ionization of the latter ($5 \cdot 10^{-7}$ cm^{-3} sec^{-1}), as well as the capture radius (equal to 10 Å) and exciton diffusion coefficient (~1 cm^2/sec).

These results agree with the data obtained in References 10 through 12. The difference between the exciton capture rate constants is small and may be caused by the dependence of the interaction cross-section of excitons with benzene molecules on temperature. When benzene is dissolved in xenon, a shallow trap for an exciton forms and the exciton capture by a benzene molecule most likely proceeds in two stages: first, the exciton is captured by a shallow trap and then from this local state the exciton is captured by a benzene molecule.

The capture effectiveness of a localized exciton by the molecule is determined by the time of its stay in a shallow trap which is exponentially dependent on the trap depth. Shallow traps may play a part at sufficiently low temperatures, and with increasing temperature the exciton capture by molecules will proceed from much deeper traps. The estimation of a trap depth is ~10^{-2} eV.[12]

Starting from data on the radiolysis of hydrocarbon in inert matrices within the exciton mechanism of radicals formation, it is possible to obtain the value of an exciton capture radius. For example, for the system of xenon-3-methylpentane at 10 K, the capture radius is of the order of 10 Å.[17]

3.2.2 Radiation-Chemical Reactions of Inorganic Molecules in the Matrix

Chemical conversions of inorganic molecules in noble gas matrices have also been studied. Solid matrices of argon, xenon, and krypton containing molecules CO_2 and NO_2 were irradiated with high-energy electrons.[24] It was shown that we are also faced with radiation energy absorption by the matrix and its transfer to impurity.

It is known that the effectiveness of the energy transfer from matrix to impurity is characterized by large values of the decay radiation yield in recalculation for radiation absorption by molecules of impurity. So, for example, the radiation yield of CO_2 decay in xenon (dissolution 1:200) is equal to 2300, and that of NO_2 in krypton (dissolution 1:400) is 1000. Treating the dependencies of the amount of dissociated molecules on the absorbed dose rate under different dilutions by the technique proposed in Reference 10, it is possible to determine $k_{ex}\tau_{ex}$ (here, k_{ex} is the exciton capture rate constant by the impurity molecule, τ_{ex} is the lifetime of the exciton). This product is equal to $4.4 \cdot 10^{-20}$ cm^3, $1.3 \cdot 10^{-19}$ cm^3, $5.9 \cdot 10^{-20}$ cm^3 for the impurity molecule NO_2 in argon, krypton, and xenon, respectively, and $2.7 \cdot 10^{-20}$ cm^3 and $10.7 \cdot 10^{-20}$ cm^3 for CO_2 in krypton and argon. It should be noted that for methane in xenon, this value is equal to $4 \cdot 10^{-21}$ cm^3. This testifies to the existence of large exciton capture radius by inorganic molecules.

3.2.3 Radiation-Chemical Reactions in Individual Substances and Their Mixtures

Study of the primary products of radiolysis in systems consisting entirely of molecules which may undergo a chemical transformation as they are irradiated, is much more difficult. Even a decrease in temperature in such systems cannot exclude secondary reactions. For example, hydrogen atoms do not become stable in alkanes except for methane, even at 4.2 K, and enter into a disproportionate reaction by a tunneling mechanism.[13,25] Hence, in spite of a large number of studies on the radiolysis of organic substances in the solid phase, the exciton mechanism of formation of radiolysis primary products may be assumed to be demonstrated only in some works.

The most convenient object for studying the primary processes of radiolysis is methane. It was shown that under γ-irradiation of conventional or deuterated methane at 4.2 K the concentration of stabilized methyl radicals and hydrogen atoms (deuterium) is similar.[26-30] In this case a hydrogen atom (deuterium) becomes stabilized near "the genus" methyl radical.[27,28] The most probable distance is 16 Å which corresponds to three constants of a methane lattice (6.76 Å). The radical pairs $CH_3 \cdots CH_3$ form in the amount of less than 1% perhaps due to a rapid decomposition of the excited complex.[31] Thus, it may be stated that a monomolecular decay of excited methane by means of predissociation mechanism from the triplet state (i.e., under localization of the triplet exciton) takes place.[2] The energy level from which the decay occurs is larger than 4.5 eV (the binding energy of C–H methane) but less than 7.5 eV, since otherwise the hydrogen atom would enter into a disproportionate reaction with methane.[32] Methane energy of the triplet state falls into this range. The triplet excitons most probably form under charge recombination, because the mobility of electrons and the rate constant of ion electron recombination in methane are sufficiently large.[34] This conclusion is also proved by the fact that in the matrix SF_6 the ions CH_4^+ were not observed even at 4.2 K.[6]

On the other hand, under radiolysis of the mixture CH_4–C_2H_6 (2 mol%) at 4.2 K, formation of only hydrogen atoms and methyl radicals is observed. Formation of ethyl radicals takes place only with the increasing temperature of a previously heated sample. It is necessary that hydrogen atoms become mobile and react with ethane molecules. Hence, the energy transfer from matrix (CH_4) to impurity (C_2H_6) does not occur. However, for a molecular crystal of methane the rate of vibrational energy relaxation in contrast to the crystals of multi-atomic molecules is much less (for methane crystals, it can reach 10^{-2} sec) and is noticeably dependent on the vibrational energy stored.[35] The dissociation of methane molecules in the solid phase is likely to happen from the uppermost triplet levels.

The dissociation rate constant by dissociation mechanism is greatly associated with the energy of the triplet state (E_T) and is as large as 10^{13} sec^{-1} at $E_T = 8.5$ eV.[35] These specific features allow the suggestion that the energy migration in solid methane is prevented by a very rapid process of localization and diactivation of the triplet excitation energy.

There also was no observation of the energy transfer from neopentane to ethane, propane, isobutane, and isobutene under radiolysis of these systems at 4.2 and 77 K.[37,38] The electron acceptors SF_6, CH_3I (SF_6^- and CH_3 radicals are observed), and the energies CCl_4 (CCl_3 radicals are observed) greatly affect the yield of the neopentane radical. The hole acceptors, tetramethylethylene, influence less.[38] This demonstrates that there is energy migration in neopentane. In fact, under radiolysis of the mixture of neopentane with cyclopentane and cyclohexane, the energy transfer takes place from neopentane to admixture, with the distance of transfer being 40 and 85 nm, respectively.[39] The absence of energy transfer from neopentane to C_2H_6, C_3H_8, isobutane, and isobutene may be explained by the fact that their triplet states are located higher than in neopentane.

Apart from this, in a number of alkanes under radiolysis of mixtures (cyclopentane, 2,3-dimethylbutane, 3-methylpentane), a transfer of energy to toluene takes place.[40-43] In this case, both the toluene luminescence and decreasing radiation yield of hydrogen are observed. The energy transfer proceeds by the exciton mechanism, the diffusion length of excitons being several hundreds of nanometers.

Interesting results were obtained under radiolysis of isobutane at 77 K. It was found that the electron acceptors SF_6, CH_3I, NO_2, and others do not affect the yield of isobutyl radicals, while the energy acceptor CCl_4 suppresses formation of the isobutyl radical.[44,45] In this case, CCl_3 radicals form. On the basis of the concentration dependence of CCl_4 and influence on the yield of isobutyl radical and other radiolysis products, a conclusion was made that there is a transfer of energy from isobutane to CCl_4 by the exciton mechanism. The capture rate constant of exciton CCl_4 was equal to $7 \cdot 10^{12}$ $m^{-1}sec^{-1}$. In addition, the data on the energy transfer by the exciton mechanism from isobutane to toluene are available. A similar result was observed in using a methylcyclohexane molecule as an impurity. The diffusion length of exciton in isobuthane was of the order of 7 to 8 nm.[39]

Energy transfer under radiolysis was observed not only in crystals but also in glasses. For example, in the matrices of the glassy isobuthane the exciton transfer of the energy from the matrix to the CCl_4 molecule takes place.[46]

The exciton mechanism of energy transfer has been studied extensively in aromatic hydrocarbon. In these substances, the electron structure is well known,[47,48] the exciton dynamics are studied,[49] and much attention has been paid to photochemistry.[50]

An experimental demonstration of energy transfer under radiolysis of aromatic substances is radioluminescence and decomposition of additions (impurities). The energy transfer under radiolysis of aromatic substances has been observed repeatedly and is beyond doubt.[51-53]

Using the methods of subnanosecond pulsed radiolysis and doublephoton laser photolysis, a mixture of toluene with various additions (chloroform, 9,10-diphenylanthracene) at 193 K was studied. It was shown that the transfer of energy to the addition occurs from the upper excitation states of toluene.[54,55]

In other studies,[56-61] radiation yields of the excitation states of 2,5-diphenyl-1,3,4-oxidiazol, naphthalene, and benzene in various matrices (including toluene) were measured under X-ray irradiation at 77 K. It was found that in toluene the excitation states of additions are formed as a result of the migration of singlet and triplet excitons of the matrix. The diffusion lengths of the transfer of a singlet and triplet exciton in toluene are 40 to 50 and 80 to 90 Å, respectively. The energy transfer on a luminescence addition is used in highly-sensitive scintillators for detecting luminescence radiations.[62]

Energy transfer in aromatic substances bringing about decomposition of the addition has been investigated less frequently. Under irradiation of solid solutions of naphthalene in polymethylmetacrylate, a photodestruction of polymer was observed as a result of energy transfer from the triplet level of naphthalene to macromolecule.[63]

The radiation yield of $FeCl_2$ at low-temperature radiolysis of ferum chloride dissolved in benzene increases up to 25.[64] The migration distance of the exciton is 80 to 100 Å. The measurements of $G_{(H_2)}$ in solid diphenyl, diphenylnaphthalene, and a mixture of diphenyl-diphenylnaphthalene show that the energy migration proceeds by the exciton mechanism. A formation of radical pairs in crystalline naphthalene, anthracene, and fluorene at 77 K is likely to proceed by the energy transfer mechanism.[65,66]

Aromatic substances in the condensed phase are characteristic of a high radiation resistance. This is due to the fact that usually the energy of the lower triplet level of these compounds is less than the breakdown energy of the C–H bond.[2] And since the energy dissipation rate noticeably exceeds the dissociation rate, the molecules not managing to dissociate find themselves in a nonreacting state. The aromatic molecules are effective captors of excitons; therefore, they are often utilized as stabilizing additions.

A study of radioluminescence at 77 K under γ-radiolysis of naphthalene solutions in hydrocarbon glasses showed that an effective transfer of energy from matrix to naphthalene takes place, and the direct excitation of naphthalene is insufficient.[67,68]

When a mixture of cyclohexane and benzene is irradiated at 77 K, the benzene is the effective defender of cyclohexane being decomposed.[69] The distance of energy transfer in such systems with use of multilayer techniques for obtaining sandwich structures was found to be 2 to 20 nm.

3.2.4 Isotope Effect in the Reactions of Excited Molecules

Interest in the reactions of excited molecules is due to the isotope effect. It should be noted that in spite of a significant number of papers which investigated the isotope effect, only a very small fraction of experimental results can be interpreted unambiguously, and it is even more difficult to indicate the relation of isotope effect to energy transfer.

For example, under radiolysis of the mixture of n-decane-h_{22} and n-decane-d_{22} a sufficient isotope effect of the radiation yield of radicals and molecules was discovered.[70-73] In References 70 and 71, this effect was explained on the basis of the exciton energy transfer mechanism, while in References 72 and 73, by selectivity of the H atom abstraction reaction. On the basis of the measurements of the radiation yield of decane dimers, the isotope effect was explained by use of ion-molecular reaction mechanism.[74]

The explanation of isotope effect by selectivity of the H atom abstraction reaction is more convincing. Papers confirm that the abstraction rate constant of hydrogen, deuterium, and tritium atoms from hydrocarbon molecules by atoms of hydrogen, deuterium, and tritium noticeably decrease in this series.[75-79] The different effectiveness of isotope decay may be caused by both differences in the effectiveness of excitons capture by isotopes[49] and different effectiveness of molecule dissociation.[2]

Of interest are studies which discovered the temperature dependence of the isotope effect in the photolysis products of the mixture of acetaldehyde and deuterated acetaldehyde, with the maximum at 27 K reaching 40.[80] The authors explain their results by the difference in the cross-section of excitons capture by acetaldehyde isotopes. Unfortunately, it remains unclear as to what bond fracture the effect is referred, as the observation was conducted for the secondary products of photolysis. Under radiolysis in the matrices of noble gases and methane and in conditions where H atoms are stabilized, the isotope effect of the primary products of radiolysis was investigated.

Under γ-radiolysis of a solid conventional and deuterated methane, it was found that $G_{(CH_3)} = G_{(H)} = 2.7$ and $G_{(G)} = G_{(CD_3)} = 2.5$. Such a difference was explained by the decreasing dissociation probability of a deuterated molecule due to its vibrational levels displacement.[81]

Under γ-radiolysis of a xenon matrix containing CH_4 or CD_4 it was demonstrated that the exciton capture rate constant by CH_4 and CD_4 molecules is equal to 2.3.[82] For CH_3 and CD_3 radicals, the constant ratio is equal to 2.4. A decrease in the capture rate constant by deuterated methane as compared to CH_4 is due to the smaller frequency of the local mode (see Chapter 5) in CD_4 and CD_3 in comparison with CH_2 and CH_3. The probability of exciton capture by the molecule (deep trap) is determined by the possibility of the dissipation of the exciton-trap energy difference (the triplet level of a molecule). In the case of CD_4 and CH_3, a larger amount of phonons is required to compensate for the energy defect, which leads to less capture probability. The techniques of the experiments used in these works do not allow judgment of the importance of the above isotope effect mechanism to be verified, since the total effect is registered.

3.2.5 The Intramolecular Transfer of Energy

The intramolecular transfer of energy plays a noticeable part in radiolysis. The migration of energy over a molecule and its localization on certain groups of a molecule greatly determine the regularities of radiolysis.

It is known that the effective intramolecular energy transfer proceeds in complicated organic molecules, for example, in polymers and biomolecules.[83] Calculations show that the excitation energy may migrate along the aliphatic chain at a distance up to 100 bonds C–C.[84] A molecule excitation may be presented by a superposition of exciton states. The localization of excitation on a C–H bond leads to its fracture.[85] The energy absorbed by the aliphatic fragment of a molecule may be transferred to the aromatic fraction which increases the photo and radiation resistance of the alkyl fraction of the molecule. Such processes were observed in the molecules of alkylbenzene[86-88] and alkyldiphenyls.[51,86] The effectiveness of energy transfer in alkylbenzene is approximately constant for the length of an aliphatic chain not exceeding four H atoms. It decreases with a further increase in the number of atoms.[88]

In radiolysis of octyl-benzene, the intramolecular transfer of energy plays a more important part compared with the intermolecular transfer.[87] The effective intramolecular transfer of energy was also observed under radiolysis of heterocycle compounds, crown-ethers.[89] The limiting stage of energy transfer in polymers is the migration of electron excitation energy to polymer.[90,91] So, in poly-N-vinylcarbazole the energy of the single excitation passes 20 nm before being localized.[92] The length which the triplet excitation energy passes in copolymers 1-vinyl naphthalene with styrene is much longer, the distance being on the order of 140 styrene chains.[93]

3.3 RADICAL REACTIONS

Significant attention in this book is devoted to solid-phase radical reactions. This is because of the simplicity and reliability of the techniques of their study and the great role they play in a radiation-chemical process. To investigate radical reactions, different ESR modifications and optical spectroscopy in infrared, visible, and ultraviolet regions are utilized. Use is also made of the importance of temperature or, to be more precise, the fact that in different conditions and temperature ranges different stages of radiolysis are effective. These and other problems are discussed in Chapters 4 to 8. Here, we shall only dwell on the common issues of chemical process development under radiation. After the physical stages of radiolysis are concluded (see Chapter 2), a sufficient amount of chemically active particles may be stabilized in a substance. Due to the fact that the absorption of energy under substance irradiation with ionizing radiation takes place nonuniformly in volume, the same nonuniformity will apply to distribution of the stabilized particles. Therefore, when considering the processes where the concentration of intermediate products is the important factor, account should be taken for the difference of the local concentration from that of the mean by sample.

There are several reactions when radicals form. Among these are ion-molecular reactions in such compounds as amines, water, methane, olefins, alcohols, and ethers. We shall only note that the mechanism of such ion-molecular processes is not always understood (see Section 5.4). Another reaction giving rise to formation of radicals is the dissociative capture of electron by molecule:

$$M + e \rightarrow A^- + R.$$

As a result, a nonparamagnetic ion and radical are generated. Finally, radicals form under dissociation of excited molecules (see Sections 1.5 and 3.2).

As a result of these reactions, radicals are accumulated in the sample and may participate in further transformations. First of all, there are reactions with the neighbouring molecules (Section 6.2). In fact, in the solid phase, abstraction reactions proceed successfully in various ranges of temperatures. It is essential to note here that the rate constant of the solid-phase reaction, especially at low temperatures, will be noticeable only if the reaction is nonendothermic. As an example, one may transfer an H atom from the alcohol matrix molecule to the methyl radical. In spite of its likely simplicity, this is a reaction with sufficiently complex nonexponential kinetics and non-Arrhenius dependence of the rate constant on temperature.

Such a known reaction as a recombination of free radicals in the solid phase has no less complicated kinetics (see Section 6.4). A radical concentration at constant temperature first rapidly decreases up to a certain value and then it practically remains unchanged even at sufficiently long observations. With increasing temperature it all repeats again.

A number of models allow some specific features of the solid-phase recombination to be explained. First is the avalanche recombination in a thermal wave where the energy, released under recombination of two radicals, propagates over a sample and stimulates a radical recombination in another part of the matrix.[94-96] At the same time, it is clear that it is impossible to explain the rather slow process of the spontaneous staircase recombination by the rapid propagation of heat over the sample. Second, radicals having a characteristic "capture sphere" for specific conditions of recombination.[97-99] Unfortunately, the mechanism by which the given kind of radicals corresponds to a "capture sphere" radius is not understood. Third, the recombination process proceeds in two stages: a diffusion approach of radicals and a recombination on reaching a critical distance between them.[100]

All of these models were supplemented with various assumptions: activation energy distribution in a recombination process over some range, influence of the medium phase state, and dependence of the recombination rate constant on radical-to-radical distance.[101-104] Interpretation of experimental data on the solid-phase recombination of radicals accounting for the migration possibility of free valences by the tunneling mechanism is presented in Section 6.4.

We shall also note an important radiation process, the conversion of radical pairs. Under the radiation effect on solid substances, the spatial distribution of radicals, as a rule, is nonuniform. Sometimes it may present closely located radical pairs. Owing to a strong exchange and dipole-dipole interaction of unpaired electrons, radical pairs have a characteristic splitting of the ESR spectrum. This splitting depends on two parameters: orientation of the pair in the magnetic field of the spectrometer and, especially important, on the distance between unpaired spins in the radical pair. It is this latter property of the radical pairs that allows the kinetics of low temperature reactions with an H-atom transfer to be investigated (Sections 4.3 and 5.5). The observation of radical pair transformation in irradiated dimethylglyoxene allowed the tunneling transfer of a hydrogen atom to be discovered. Another interesting process of H-atom transfer with a formation of radical pairs was observed in a fluorene-acridine system. In this case, excitation of acridine molecules led to formation of a radical pair.

Besides the spontaneous radical reactions and processes occurring under light absorption by radicals (see Chapter 7), it is necessary to take into account the possibility of radical excitation by energy which originates in γ-radiation. First are excitons which may cause the decay of radicals (see Section 6.1). The observed increase in the ratio of H-atom radiation yield to the yield of radicals,[17,18] a constant yield of molecular hydrogen, and a decreasing yield of radicals under an increasing radiation dose rate[10] may be explained by the decay of radicals in the course of interaction with excitons of the inert gas matrix.[105]

The energy of the primary excitation state and the ionization potential of a radical, as a rule, are less than those of the corresponding molecules so that radicals are the effective capturers of excitons. They may capture both triplet and singlet excitons. This is due to the fact that forbidden transfers may be realized in the presence of paramagnetic acceptors.

It was possible to simulate this process experimentally. Under photolysis of a number of radicals in a xenon matrix at 5 K,[106,107] the decay of radicals took place with a simultaneous increase in the concentration of stabilized H atoms. The results favor the monomolecular decay mechanism. The following radical reactions were observed:

$$CH_3 + \hbar\nu\ (185\ nm) \rightarrow CH_2 + H$$

$$C_2H_5 + \hbar\nu\ (185\ nm) \rightarrow C_2H_4 + H$$

$$C_2H_3 + \hbar\nu\ (185\ nm) \rightarrow C_2H_2 + H$$

$$HCO + \hbar\nu\ (500\ nm) \rightarrow H + CO$$

$$HO_2 + \hbar\nu\ (254\ nm) \rightarrow H + O_2$$

$$CH_3O_2 + \hbar\nu\ (254\ nm) \rightarrow CH_3 + O_2$$

Besides, the biradical CH_2 reacted with a methyl radical:

$$CH_2 + CH_3 \rightarrow C_2H_5^* \rightarrow C_2H_5$$

$$CH_2 + CH_3 \rightarrow C_2H_5^* \rightarrow C_2H_3 + H_2$$

and biradical CH_2:

$$CH_2 + CH_2 \rightarrow C_2H_4^* \rightarrow C_2H_3 + H.$$

These reactions proceeded only at temperatures above 35 K in methane and 70 K in xenon when the CH_2 radical became mobile. In addition, there was a decrease in the concentration of methyl radicals when an increased absorbed energy dose rate in a xenon matrix containing methane was observed. It is explained by the decay of radicals under its interaction with excitons in xenon. The rate constant of the exciton matrix capture by a radical was $6.5 \cdot 10^{-9}$ cm^{3}sec^{-1} which is essentially higher than the capture rate constant by methane, $8.4 \cdot 10^{-10}$ cm^{3}sec^{-1}. Thus, a methyl radical is a more effective trap than a methane molecule.

A sensitization of the radiation decay of stable radicals by the matrix takes place.[108] Under radiolysis of a stable nitric radical (the concentration being $2.4 \cdot 10^{19}$ to $2.5 \cdot 10^{20}$ radicals per gram) in isopropyl alcohol, the energy is transferred from matrix to impurity which leads to a high radiation yield of radical decay (240 radicals per 10 eV of the energy absorbed).

In addition, the absorption of excitons by radicals results in a decreasing radiation yield of radicals (if radicals form under dissociation of excited molecules) due to decreasing exciton lifetime. It is clear that with increasing concentration of excitons the effectiveness of energy dissipation in a medium increases. Thus, if the exciton-radical chain does not develop, then radicals may cause a stabilizing effect on the radiolysis process (autostabilization). So, the triplet excitons of anthracene interact with radicals produced under X-ray radiation.[109,110] The lifetime of triplet excitons decreased with increasing absorbed dose rate. Transformations of radicals in these works were not studied. Also, a study was undertaken of extinguishing singlet excitons by defects formed under irradiation in anthracene with electrons.[111] At the same time it should be pointed out that, with small absorbed dose rates and when the radicals' concentration is still insufficient to extinguish excitons, the stabilizing effect of radicals is insignificant.

A radical is also an effective acceptor of holes. As a result of the capture of a hole, the radical transforms into a nonparamagnetic positively charged particle, and under further recombination with an electron the excitation of a radical occurs, with development of a chain reaction being possible, as in the case of exciton absorption. The capture of holes by radicals may also cause a stabilizing effect on the process of radiolysis.

Radical reactions may be initiated under absorption of Cerenkov radiation. It was shown that the Cerenkov radiation arising when high-speed electrons pass through a substance can make a noticeable contribution to a radiation-chemical process.[112] Electrons with the energy 10 MeV at the flux density 10^{12} to 10^{14} $cm^{-2}sec^{-1}$ yield the photon flux 10^{15} to 10^{17} $cm^{-2}sec^{-1}$ in the range of 200 to 600 nm, which corresponds to a flux in conventional electric bulbs used in photochemical studies.

Utilization of magnetic effects is considered very promising for the study and control of radical processes.[113,114] In the case of interaction of triplet excitons and radicals,[115] the effect is contained in the action of a magnetic field in a spin state of a short-lived intermediate complex. The probability of different ways for the intermediate complex to decay depends on its spin state which is changed by field effect. In addition, the discovery of magnetic effects helps in some cases to make a choice between the processes of energy transfer and charge transfer.[116] It was shown that, under interaction of triplet tetracene excitons with oxygen molecules, energy transfer through a formation of intermediate short-lived complexes takes place.[117] An increase in the magnetic field strength leads to an increase in the capture rate constant of exciton by oxygen molecule.

3.4 PROCESSES WITH PARTICIPATION OF STABILIZED ELECTRONS

The effectiveness of chemical transformations of ions formed under substance irradiation depends on the lifetime of an electron-cation pair with respect to recombination. In condensed media, the importance of heminal recombination (recombination of an electron with the parent ion) increases. Its probability is determined by a ratio between the Onzager radius ($r_0 = e^-e^+/\varepsilon\, k\, T$; e^- and e^+ are the charges of the electron and cation, respectively; ε is the static dielectric permeability of the medium) and the distance r_T at which the electron is thermalized, e.g., its kinetic energy decreases up to thermal. If for the considered substance and type of radiation $r_T > r_0$, then the electron can avoid a rapid recombination by the parent ion and participate in reactions with molecules of the substance or may be captured by a trap (other ions, radicals, lattice defects). Consequently, the yield of separated ion pairs in a substance depends on its dielectric permeability, probability of electron thermalization, concentration of electron traps, and types of radiation.

With large local concentrations of cations and when the distances between them do not exceed r_T, the probability of a heminal recombination is of the same order as the probability of electron recombination with neighbouring ions. Thus, the direction of the chemical reaction and the yield of radiolysis products depend on the local concentration of active particles and medium properties.

It is possible to separate several principal processes by participation of thermalized electrons. First of all is a cation recombination with formation of excited molecules which can dissociate (see Section 1.5). Apart from this, an electron is captured by radicals and molecules with positive electron affinity. Naturally such a capture may be dissociative which gives rise to further chemical transformations. In addition, after being thermalized, the electron participates in the physical stages of radiolysis. Interacting with a hole, the electron forms an exciton (see Section 1.1), and its capture by the structural defects of the lattice leads to the appearance of dye centers. The effectiveness of some processes with participation of an electron is greatly determined by the dielectric permeability of the medium. It should be pointed out that the radiation-induced elementary excitations can be very active chemically and participate in irreversible transformations of substance.

REFERENCES

1. Basilevskii, M.V., Gerasimov, G.N., Petrochenko, S.I., Tikhomirov, V.A., Quantum-chemical investigation of the mechanism of ionic polymerization in crystals, *Chem. Phys.,* 55, 259, 1981.
2. Plotnikov, V.G., Ovchinnikov, A.A., Photo- and radiation chemical molecule stability. Monomolecular reaction of hydrogen atom abstraction, *Russ. Chem. Rev.,* 47, 444, 1978.
3. Plotnikov, V.G., Regularities of the processes of radiationless conversion in polyatomic molecules, *Intl. J. Quant. Chem.,* 16, 527, 1979.
4. Andrews, L., Prochaska, F.T., Ault, B.S., Matrix photoionization and radiolysis of CH_2Cl_2 and CH_2Br_2 infrared and ultraviolet absorbtion spectra and photolysis of CH_2Cl_2 and $CH2B^+_2$, *J. Amer. Chem. Soc.,* 101, 9, 1979.
5. Arai, S., Kira, A., Imamura, M., Mechanism of formation of cationic species in the radiolysis of butyl chlorides, *J. Phys. Chem.,* 80, 1968, 1976.
6. Toriyama, K., Nunone, K., Iwasaki, M., Structures and reactions of radical cations of some prototype alkanes in low temperature solid as studied by ESR spectroscopy, *J. Phys. Chem.,* 77, 5891, 1982.
7. Kaplan, I.G., On the distribution of ionizing radiation energy in molecular mixtures, *Dokl. Phys. Chem.,* 164, 842, 1965.
8. Kaplan, I.G., Popova, L.V., Khadzhibekova, L.M., The calculation of degradation spectrum of fast electrons in continuous homogeneous medium, *High Energy Chem.,* 7, 241, 1973.
9. Grigoriev, E.I., Psheezhetskii, S.Ya., Trakhtenberg, L.I., Exciton mechanism of radical format in Xe matrix on bombardment by metastable Xe and Ar atoms of surface, *High Energy Chem.,* 19, 41, 1985.
10. Grigoriev, E.I., Psheezhetskii, S.Ya., Slavinskaya, N.A., Trakhtenberg, L.I., Mechanism of radicals formation on the irradiation on xenon CH_4 containing matrix, *High Energy Chem.,* 21, 519, 1987.
11. Fugol, I. Ya., Exciton in rare-gas crystals, *Adv. Phys.,* 27, 1, 1978.
12. Grigoriev, E.I., Psheezhetskii, S.Ya., Slavinskaya, N.A., Trakhtenberg, L.I., Stabilizing effect of aromatic additives in the exciton mechanism of formation of radicals, *High Energy Chem.,* 22, 31, 11988.
13. *Rare Gas Solid,* Klein, M.L., Venabls, J.L., Eds., Academic Press, New York, 1977.

14. Bouldin, W.V., Patten, R.A., Gordy, W., Energy migration and transfer in solid argon and krypton and low temperatures, *Phys. Rev. Lett.,* 9, 98, 1962.
15. Bouldin, W.V., Gordy, W., Energy migration and isotopic effect in irradiation solids at low temperature, *Phys. Rev.,* 135, 806, 1964.
16. Florin, R.E., Brown, D.W., Wall, L.R., g-Irradiation of small molecules at 4 and 77 K, *J. Phys. Chem.,* 66, 2672, 1962.
17. Bhattachaya, D., Willard, J.E., Radiolysis productions of trapped hydrogen atoms from organic compounds in xenon, krypton, *J. Phys. Chem.,* 85, 154, 1981.
18. Muto, H., Toriyama, K., Numone, K., Iwasaki, M., Radiolysis of alkanes and olefines in xenone matrices at 4.2 K as studied by ESR formation and trapping by hydrogen atoms and their subsequent reactions at cryogenic temperatures, *Rad. Phys. Chem.,* 19, 201, 1982.
19. Ausloos, P., Robbert, R.E., Lias, Sharon, G.J., Direct and inert-gas-sensitized radiolysis and photolysis of methane in the solid phase, *J. Chem. Phys.,* 42, 540, 1965.
20. Scala, P.A., Ausloos, P., Condensed-phase photolysis and radiolysis of 2-methylbutane, *J. Chem. Phys.,* 47, 5129, 1967.
21. Ophir, Z., Raz, B., Jortner, J., Exciton-enhanced photoemission from doped solid rare gases, *Phys. Rev. Lett.,* 33, 415, 1974.
22. Ophir, Z., Raz, B., Jortner, J., Photoemission from doped solid rare gases, *J. Chem. Phys.,* 62, 650, 1975.
23. Schwentner, N., Koch, F.F. Ophir, Z., Jortner, J., Photoelectron energy distribution measurement from benzene in rare gas matrix, *Chem. Phys.,* 34, 281, 1978.
24. Pacansky, J., Maier, M., Irradiation of small molecules isolated in rare gas matrices by high energy electron beams, *J. Mol. Struct.,* 222, 33, 1990.
25. Wang, H.Y., Willard, J.E., Effects of molecular structure and temperature on the tunneling abstraction of deuterium from carbon-deuterium bonds by trapped hydrogen atoms and their subsequent reactions of cryogenic temperatures, *J. Phys. Chem.,* 83, 2585, 1979.
26. Smaller, B., Matheson, M.S., Paramagnetic species produced by g-irradiation of organic compounds, *J. Chem. Phys.,* 28, 1169, 1958.
27. Gordy, W., Morehouse, R., Triplet-state spin resonance of an H-atom-methyl-radical complex in a solid matrix, *Phys. Rev.,* 151, 207, 1966.
28. Toriyama, K., Iwasaki, M., Nunone, K., ESR studies of irradiated methane and ethane at 4.2 K and mechanism of pairwise trapping of radicals in irradiated alkanes, *J. Chem. Phys.,* 71, 1698, 1979.
29. Timm, D., Willard, J.E., Absence of hydrogen atom production in radiolysis of solid hydrocarbons, *J. Phys. Chem.,* 73, 2403, 1969.
30. Iwasaki, M., Toriyama, K., Muto, M., Nunone, K., Hydrogen abstraction by thermal hydrogen atoms at 10 to 30 K. Electron spin resonance evidence, *Chem. Phys. Lett.,* 56, 494, 1978.
31. Jakimchenko, O.E., Lebedev, Ya.S., Radical pairs in study of elementary chemical reactions in solid organic substances, *Russ. Chem. Rev.,* 47, 1018, 1978.
32. Wakahara, A., Miyazaki, T., Fueki, K., Effects of energy of hydrogen atom on trapping and reaction in solid methane at 4 K, *J. Phys. Chem.,* 86, 1781, 1982.
33. Kunz, A.B., Electron structure and optical properties of solid methane, *Phys. Rev. B,* 28, 3465, 1983.
34. Nakayama, Y., Shinsuka, K., Natano, Y., Electron mobilities and electron-ion recombination rate constants in solid, liquid and gas, *J. Chem. Phys.,* 78, 5820, 1983.
35. Ovchinnikov, A.A, Localised long-lived vibrational states in molecular crystals, *Sov. Phys. JETF,* 57, 263, 1969.
36. Kinugawa, K., Miyazaki, T., Hydrogen atom abstraction reaction H-atom in g-radiolysis and U,V-photolysis of alkane mixtures at 4 K, *Rad. Phys. Chem.,* 20, 341, 1977.

37. Iwasaki, M., Toriyama, K., Nunone, K., Fukaya, M., Muto, M., Effects of irradiated temperature of hydrogen atom reaction in neopentane, *J. Phys. Chem.,* 81, 1410, 1977.
38. Iwasaki, M., Muto, M., Toriyama, K., Fukaya, M., Nunone, K., Irradiation temperature effects of hydrogen atom reactions upon radiolysis and photolysis at 4.2 K and 77 K of neopentane containing various additives as studied by electron spin resonance, *J. Phys. Chem.,* 83, 1590, 1979.
39. Miyazaki, T., The migration of the exciton in the radiolysis of alkanes, *Bull. Chem. Soc. Jap.,* 46, 329, 1973.
40. Miyazaki, T., Yammamoto, M., Energy transfer in the radiolysis of solid hydrocarbons at 77 K: mono-olefins, *Rad. Phys. Chem.,* 10, 247, 1977.
41. Tanaka, T., Miyazaki, T., Kuri, Z., Formation of excited singlet toluene in the radiolysis of alkane containing toluene in the solid phase at 77 K, *Rad. Phys. Chem.,* 8, 645, 1976.
42. Miyazaki, T., Saitake, Y., Kuri, Z., Sakai, S., Luminescence from 2,3-dimethylbutane containing toluene in the solid phase at 77 K during g-irradiation, *Bull. Chem. Soc. Jap.,* 46, 70, 1973.
43. Miyazaki, T., Excitation transfer in the radiolysis of alkanes in the solid phase at 77 K, *Rad. Phys. Chem.,* 8, 57, 1976.
44. Wakayama, T., Miyazaki, T., Fueki, K., Kuri, Z., Radical formation in the radiolysis of solid isobutane as studied by ESR spectroscopy, *Bull. Chem. Soc. Jap.,* 42, 1164, 1969.
45. Wakayama, T., Miyazaki, T., Fueki, K., Kuri, Z., Excitation transfer in the radiolysis of solid hydrocarbons as studied by ESR spectroscopy and product analysis, *Bull. Chem. Soc. Jap.,* 43, 1017, 1970.
46. Miyazaki, T., Yamada, T., Wakayama, T., Fueki, K., Kuri, Z., Effect of phase on excitation transfer in radiolysis, *Bull. Chem. Soc. Jap.,* 44, 934, 1971.
47. Davydov, A.S., *Theory of Molecular Excitons,* John Wiley & Sons, New York, 1961.
48. Gutmann, F., Lyons, LF., *Organic Semiconductors,* John Wiley & Sons, New York, 1967.
49. Agranovich, V.M., Galanin, M.D., *Energy Transfer of Electron Excitation in Condensed Matter,* Nauka, Moscow, 1978 (in Russian).
50. Birks, J.B., *Photophysical Processes of Aromatic Compounds,* John Wiley & Sons, New York, 1970.
51. *Radiation Chemistry of Hydrocarbons,* Foldiak, G., Ed., Elsevier, Amsterdam, 1981.
52. Gäumann, T., Hoigne, J., Eds., *Aspects of Hydrocarbon Radiolysis,* Academic Press, New York, 1968.
53. Saraeva, V.V., *Radiolysis Hydrocarbons in Liquid Phase,* Moscow University, Moscow, 1986 (in Russian).
54. Beck, G., Richard, J.T., Thomas, J.K., Detection of higher exited states of toluene, *Chem. Phys. Lett.,* 40, 300, 1976.
55. Beck, G., Thomas, J.K., Two photon laser photolysis of toluene, *Soc. Farad. Trans. I.,* 72, 2610, 1976.
56. Andreeshchev, E.A., Kilin, S.F., Rodman, I.M., Spurny, F., On the radioluminiscence of organic substances. VII, *Izv. Acad. Nauk SSSR. Ser. Fiz.,* 34, 662, 1970.
57. Spurny, F., Kilin, S.F., Rodman, I.M., On the radioluminiscence of organic substances. VIII, *Optics Spectrosc.,* 28, 934, 1970.
58. Andreeshchev, E.A., Kilin, S.F., Novotij, I., Rodman, I.M., Spurny, F., Excitation energy transfer in benzene and p-xylene solution, *Optics Spectrosc.,* 24, 229, 1968.
59. Spurny, F., Kilin, S.F., Rodman, I.M., Quenching fluorescence anthracene by carbon tetrachloride, *Izv. Acad. Nauk SSSR. Ser. Fiz.,* 32, 1373, 1968.

60. Spurny, F., Andreeshchev, E.A., Kilin, S.F., The radioluminiscence of organic materials. The yields of excited states in frozen solutions stimulated by X-rays, *High Energy Chem.,* 7, 11, 1973.
61. Andreeschev, E.A., Kilin, S.F., Spurny, F.I., On the mechanism of radioluminescence in alliphatic solvent, *J. Rad. Phys. Chem.,* 5, 431, 1973.
62. Pikaev, A.K., *Dosimetry in Radiation Chemistry,* Nauka, Moscow, 1975 (in Russian).
63. Batekha, I.G., Shekk, I.B., Energy transfer from high triplet levers as a cause of polymer photodestruction, *Dokl. Phys. Chem.,* 197, 614, 1971.
64. Phillips, D.H., Schug, J.C., Luminiscence from aromatic polymers, monomers and dimers under high-energy electron excitation, *J. Chem. Phys.,* 50, 3297, 1969.
65. Matsuyama, T., Yamaoka, H., Radical pairs in irradiated aromatic crystals, *J. Phys. Chem.,* 68, 331, 1978.
66. Matsuyama, T., Yamaoka, H., Radical pairs formation in some aromatic polycrystals at 77 K, *J. Chem. Phys. Lett.,* 57, 269, 1978.
67. Brockbehurst, B., Russell, R.D., Ion recombination in g-irradiated naphthalene solutions, *Trans. Farad. Soc.,* 65, 2159, 1969.
68. Badger, B., Broaclehurst, B., Formation of dimer cations of aromatic hydrocarbons, *Nature,* 219, 263, 1968.
69. Dyne, P.J., Denhartog, J., Radiolysis of solid cyclohexane-benzene mixtures at 196°C, *Nature,* 202, 1105, 1964.
70. Golobro, T., Lund, A., Energy transfer in g-irradiated $C_{10}D_{22}$–$C_{10}H_{22}$ crystals, isoptope effect of radical formation, *Chem. Phys. Lett.,* 27, 300, 1974.
71. Claesson, O., Lund, A., H/D isotope effect in g-irradiated $C_{10}D_{22}$–$C_{10}H_{22}$ mixtures studied by mass spectrometry, *Chem. Phys. Lett.,* 47, 155, 11977.
72. Miyazaki, T. et al., Selective hydrogen atom abstraction, H-atom in g-radiolysis, and photolysis at 77 K, *Bull. Chem. Soc. Jap.,* 51, 1676, 1978.
73. Gillbro, T., Lund, A., Desposition of radiation energy in solid as visualized by distribution, structure and properties of alkyl radicals in g-irradiated n-alkane single crystals, *J. Rad. Phys. Chem.,* 8, 625, 11976.
74. Tilguin, B. et al., Yields of dimeric products in g-irradiated $C_{10}H_{22}$–$C_{10}D_{22}$ mixtures at 77 K. Evidence of an H/D isotope effect, *J. Phys. Chem.,* 86, 3324, 1982.
75. Wang, H.Y., Willard, Y.E., Effects of molecular structure and temperature on the tunneling abstraction of D from C–D bonds by trapped H and D atoms in alkane glasses at 5–29 K, *J. Phys. Chem.,* 83, 2585, 1979.
76. Miyazaki, T., Wakahar, A., Kimura, K., Fueki, K., Energy phase and temperature affect hydrogen atom abstraction reaction from isobutane by H atoms at 4 and 77 K, *J. Phys. Chem.,* 85, 564, 1981.
77. Aaxono, Y., Tachikawa, E., Miyazaki, T. et al., Comparison of multiphoton ionization-fragmentation behavior of benzene, fluorbenzene and toluene, *J. Phys. Chem.,* 87, 1701, 1983.
78. Saeki, M., Tachikawa, E., Miyazaki, T. et al., Selective hydrogen atom abstraction by tritium and deuterium atoms in the radiolysis of tritiated decane at 77 K. Mass effect on solid-state reaction, *J. Phys. Chem.,* 88, 3108–32110, 1984.
79. Toriyama, K., Iwasaki, M., Tunneling effects on hydrogen abstraction by hydrogen atoms and their trapping in some hydrocarbons irradiated at low temperatures, *J. Phys. Chem.,* 82, 1056, 1978.
80. Pervukhin, V.V., Obynochny, A.A., Moralev, V.M., Sagdeev, R.Z., Temperature maximum of isotope effects in the method of isotope traps, *High Energy Chem.,* 15, 377, 1981.
81. Wilkey, D.D., Fenrick, H.W., Hager, S.L., Bremer, N., Willard, J.E., Deuterium isotope effects in the radiation chemistry of glassy and polycrystalline hydrocarbons, *J. Chem. Phys.,* 66, 1170, 1977.

82. Grigoriev, E.I., Zhukov, D.O., Zakharova, V.V., Trakhtenberg, L.I., Deuteron effect in the exciton formation and destruction of methyl radicals in solid xenon, *High Energy Chem.,* 27, 46, 1993.
83. *Radiationless transfer of the electron excited energy,* Ermolaev, V.L., Ed., Nauka, Leningrad, 1977 (in Russian).
84. Kaplan, I.G., Plotnikov, V.G., On migration of the energy electronic excitation along an aliphatic chain, *High Energy Chem.,* 1, 507, 1967.
85. Partridge, R.H., Excitation energy transfer in alkanes. I. Exciton model, *J. Chem. Phys.,* 52, 2484, 1970.
86. Jones, K.H., Van Dusen, W., Theard, L.M., Intermolecular and intramolecular energy transfer in gamma-irradiated alkylbenzenes and related mixtures, *Rad. Res.,* 23, 128, 1964.
87. Bagdasarian, Kh.S., Izrailevich, N.S., Krongauz, V.A., Intramolecular migration of energy upon the radiolysis of alkylbenzenes, *Dokl. Phys. Chem.,* 141, 887, 1961.
88. Zeman, A., Heusinger, H., Intermolecular energy transfer in irradiated alkylbenzenes, *J. Phys. Chem.,* 70, 3374, 1966.
89. Grigoriev, E.I., Myasoedova, T.G., Nesterov, S.V., Trakhtenberg, L.I., The influence of structure of crown ethers on their radiation resistance, *High Energy Chem.,* 21, 437, 1987.
90. Àlfimov, M.V., Batekha, I.G., Shoek, Yu.B., Phosphorescense sensibilized by solvent, I. *High Energy Chem.,* 2, 215, 1968.
91. Àlfimov, M.V., Batekha, I.G., Shoek, Yu.B., Solvent sensibilized phosphorescense. II. Diffusion-controlled energy transfer to the admixtures, *High Energy Chem.,* 3, 48, 1969.
92. Bauser, H., Klopfer, W., Exciton in polymers with aromatic groups, *Kolloid. Z. Z. Polymers,* 4241, 1026, 1970.
93. Fox, R.B., Driel, T.R., Cozzens, R.F., Intermolecular triplet energy transfer in poly(1-vinylnaphthalene). *J. Chem. Phys.,* 50, 1532, 1961.
94. Jackson, J., Montroll, E., Free radical statistics, *J. Chem. Phys.,* 28, 1101, 1958.
95. Jackson, J., Dynamic stability of frozen radicals. I. Description and application of the model, *J. Chem. Phys.,* 31, 154, 1959.
96. Jackson, J., Dynamic stability of frozen radicals. II. The formal theory of the model, *J. Chem. Phys.,* 31, 732, 1959.
97. Wait, T.R., Theoretical treatment of the kinetics of diffusion-limited reactions, *Phys. Rev.,* 107, 463, 1957.
98. Wait, T., General theory of bimolecular reaction rates in solids and liquids, *J. Chem. Phys.,* 28, 103, 1958.
99. Wait, T., Bimolecular reaction rates in solids and liquids, *J. Chem. Phys.,* 32, 21, 1960.
100. Lebedev, Ya.S., Kinetics of biomolecular reactions in condensed phase. Rates calculation based on Cagial model, *Kinet. Catal.,* 8, 245, 1967.
101. Bensansson, R., Durur, M., Dworkin, A., Magat, M., Marx, R., Szwarc, H., Role of molecular motions in reactions of radicals produced by irradiation and trapped in solid matrices, *Disc. Farad. Soc.,* 36, 177, 1963.
102. Smith, W.V., Jacobs, D.E., Kinetics of disappearance of radicals trapped in irradiated polyethylene, *J. Chem. Phys.,* 37, 141, 1962.
103. Cracco, F., Arvia, A.J., Dole, M., ESR studies of free radical decay in irradiated polyethylene, *J. Chem. Phys.,* 37, 2449, 1962.
104. Roch, J., La resonance paramagnetique electronique et son application a l'etude de Quelques phenomenes physico-chimiques, *Annal. de Physique.,* 5, 1401, 1960.
105. Trakhtenberg, L.I., Milikh, G.M., Kinetics of free radical growth under irradiation of solids, *High Energy Chem.,* 17, 483, 1983.

106. Bhahacharya, D., Willard, J.E., Photoelimination of H from radicals in CH_4 and Xe matrices. Photodecomposition of CH_3O_2, *J. Phys. Chem.,* 86, 962, 1982.
107. Bhahacharya, D., Willard, J.E., Reactions of CH_2 with CH_2 and CH_3 in CH_4 and Xe matrices, *J. Phys. Chem.,* 86, 967, 1982.
108. Tikhomirov, L.A., Buben, N.I., Radiolysis of solid solutions of stable radicals in isopropyl alcohol, *Dokl. Phys. Chem.,* 163, 414, 1965.
109. Ern, V., Merifields, R. Magnetic field effect on triplet excitation quenching in organic crystals, *Phys. Rev. Lett.,* 21, 609, 1968.
110. Weisz, S.Z., Richardson, P., Cobas, A., Jarnagin, R.C., Triplet sampled radiation damage, *Mol. Crystals,* 4, 277, 1968.
111. Han-Magomedova, Sh.O., Decay singlet excitons by defect in anthracene, *Trudy FIAN,* 59, 236, 1972.
112. Kaplan, I.G., Polyanskii, N.V., On the possible role of Vavilov-Cherenkov's radiation in radiation chemistry, *High Energy Chem.,* 16, 387, 1982.
113. Anisimov, O.A., Malin, Ya.N., The use of magnetic effects for primary radiation chemical processes in liquids studies, *High Energy Chem.,* 14, 307, 1980.
114. Buchachenko, A.I., Sagdeev, Z.Z., Salikhov, I.M., *Magnetic and Spin Effects in Chemical Reactions,* Nauka, Novosibirsk, 1988 (in Russian).
115. Buchachenko, A.I., in *Physical Chemistry Modern Problems,* Kolotyrkin, Ya.M., Ed., Khimia, Moscow, 1980 (in Russian).
116. Leshina, T.V., Beliaeva, S.G., Mariasova, V.J., Sagdeev, R.Z., Molin Yu. N., Application of magnetic effects for the study of the mechanism of trans-cis isomerization of stilbenes, *Dokl. Phys. Chem.,* 255, 141, 1980.
117. Francevich, E.A., Die ionisierungsquerschnitte ion stickstoff- and kohlenwasserstoff-deri-vatten, *High Energy Chem.,* 14, 195, 1980.

4 LOW-TEMPERATURE RADIATION-CHEMICAL REACTIONS

Material presented in this chapter does not present a full discussion of methods used to consider low-temperature processes; at present there are several monographs and reviews available on this issue.[1-12] Here, we shall explain only those results that allow the interpretation of numerous experimental data on the low-temperature transfer of atomic particles. The electron transfer will not be treated in this chapter. This subject is explained fairly completely in a number of reviews (see, for example, References 2, 3, 5, 8, 11, 13–15) and is discussed briefly in Chapter 3.

Since the question is about solid phase processes, then this fact already imposes a constraint on temperature, which should not be too high because many of the substances studied have low temperature melting points. At the same time, in these experimental conditions and at lower temperatures, atomic particle tunneling plays an important role to which this chapter will pay much attention.

All the processes to be discussed in this chapter proceed only after initiation by radiation. Sometimes this is a high-energy radiation, and in other cases it is optical. At the same time, the method of initiation does not influence the reaction regularities. Therefore, in order to provide a more consistent and scalene idea about the specific features of low-temperature chemical processes, we shall discuss the literature data obtained by different means of initiation.

First of all, consideration will be given to models used to describe a chemical reaction in the solid phase; the main attention is devoted to means that render it possible to account for the effect of intermolecular motions in the process of atomic transfer. It is shown that the discussion of intermolecular motions primarily leads to a description of the regularities of low-temperature reactions.

Further, a review is given of the experimental results and an explanation of special mathematical means that are required. The chemical reaction kinetics are described by these specially developed means and differ from the ordinary Arrhenius kinetics.

4.1 GENERAL KNOWLEDGE ABOUT THE PECULIARITIES OF LOW-TEMPERATURE REACTIONS

The description of the dependence of the chemical reaction rate constant on temperature (T), as well as on the medium and reagents, is based on over 100 years of the Arrhenius law:[16,17]

$$K(T) = K_0 \exp[-E_a/k_B T], \tag{4.1a}$$

where E_a is the activation energy of the chemical reaction, k_B is the Boltzman constant, K_0 is the pre-exponential factor.

According to this law only molecules with energy greater than the activation energy will enter into chemical reaction. Naturally, in conditions of heat equilibrium the number of such molecules will be proportional to the exponent from Formula (4.1).

At the present time, there are different approaches to describe an elementary act of chemical reaction. This difference manifests only in the dependence of the pre-exponential factor to temperature as well as to the properties of the medium and reagents. The exponent remains unchanged.

For instance, in accordance with a theory of transient state,[18] the expression for the rate constant has the form of Equation (4.1) in which the pre-exponential factor is equal to:

$$K_0 = K_{TST}^0 = 2\pi \frac{k_B T}{\hbar} \frac{Z^\#}{Z_0}. \tag{4.1b}$$

Here, $\hbar$ is the Planck's constant, Z_0 and $Z^\#$ are the statistical sums of the initial and transient states of the reaction system.

Formulas for the rate constant, (4.1a) and (4.1b), are derived by an assumption of complete thermodynamic equilibrium in a reaction system. In addition, the possibility of inverse transitions is neglected. To account for inverse transitions, an additional factor usually is introduced which is called a transmission coefficient.

In the theory of transient states, no account is taken of the dynamics of a small adjustment of the medium which, as a rule, accompanies a chemical reaction. The first attempt to account for the influence of medium dynamics was undertaken in the well known work by Kramers.[19] According to this approach, the elementary act is described as the generalized Brownian motion in the potential field. The rate constant keeps the former exponential form (4.1a). The pre-exponential factor in the given case is shown by expression

$$K_0 = \kappa K_{TST}^0 \tag{4.1c}$$

Here, the medium influence on the elementary act is accounted for in the explicit form through the function of the parameter κ to the coefficient of friction.

The expression for the pre-exponential factor (4.1c) was introduced earlier in Reference 20. The value of κ in the given case was considered as a transmission coefficient, as mentioned above, taking into account the possibility of unsuccessful interaction of reagents.

It should be noted that the Kramers results can be presented in the form of the rate constant derived by the theory of transient states for the extended system. This extended system includes a reaction system and oscillator reservoir equivalent to friction force and random force, which act along the reaction mode.[8,21] In several recent publications,[22-26] a method is developed which is equivalent to the Kramers approach and in which, to allow for the medium dynamics, additional effective coordinates are introduced. The motion along these coordinates describes the dynamics of the effective modes of the diluent responsible for the interaction of the diluent and reagents.

Until now the question was about the pre-exponential factor in Equation (4.1a), and the exponent (activation energy) did not depend on friction. Generally speaking, such an aspect has a constrained area of application. Recently it has been shown that friction can influence not only the pre-exponent but can also lead to alteration of the activation energy of the chemical process.[27,28]

If we talk of the temperature dependence of the exponent (4.1), the change of its shape and a change of the activation energy value can be achieved only on the basis of counting wave properties of atomic particles. As a rule, use of quantum mechanics is very productive in chemical kinetics. There are quantum-chemical calculations of the potential energy surface in a chemical reaction, spectrum discreteness of an electron, vibrational and rotary energies of molecules, and the energy of zero vibrations changing the potential barrier value of chemical reactions and determining the energy defect (ΔE) at the particle transfer (see Figure 4.1). Furthermore, it is necessary to account for the possibility of the tunneling mechanism overcoming the reaction barrier and the influence of the medium rearrangement on charge transfer process which will be considered in this chapter.

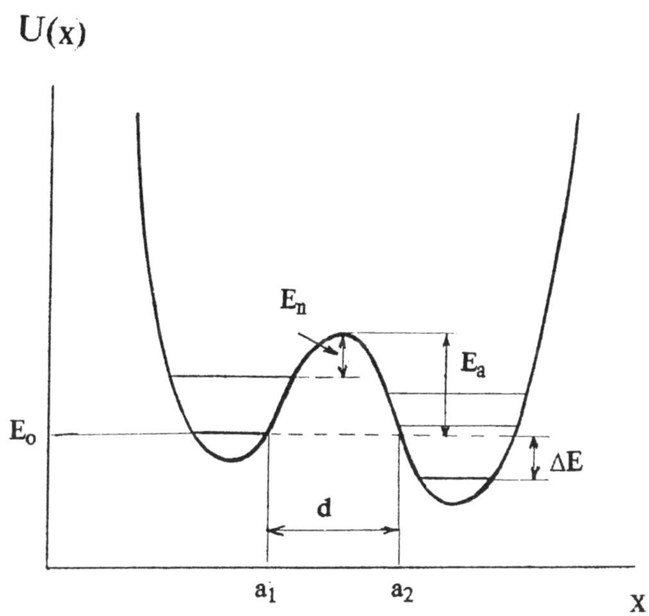

Figure 4.1 One-dimensional potential barrier for chemical reaction.

Early works on tunneling were associated with chemical processes. First, Hund considered the task of delocalization of a particle between double potential wells,[29] then Gamov's work on the α-decay of heavy nuclei appeared.[30] Only then were the articles in References 31 to 33 published, in which the possible role of tunneling in chemical transformations was analyzed.

In the 1930's, it was Bell[34-36] who obtained the most important results on tunneling as applied to chemical reactions. Later, he generalized them in References 1 and 37, which analyze the role of tunneling both in solvent and gas phase, explain possible isotope effects, and provide numerous examples of tunneling over barriers of different shapes.

A new interest in investigating the contribution of tunneling transitions to chemical transformations arose again in 1959, starting from the work by Goldanskii[38,39] in which a criterion was given of the exponential prevailance of tunneling transitions over the Arrhenius ones; above all, the importance of zero intramolecular vibrations was shown. It seems that the most significant results of these works were the appearance and later development of concepts about the tunneling low-temperature rate limit of chemical reactions.

It is interesting that, although the papers[38,39] appeared about 40 years ago, Western scientists were not drawn to them until the 1980s. In our opinion, there are several reasons for this: first (and it has been noted earlier[6]), the journal *Doklady AN SSSR* was not translated into English in 1959, and, second, experimental works on tunneling phenomena at super low temperatures stimulated a new cycle of theoretical investigations in the 1970s and 1980s which were based to some extent on References 38 and 39. Another very important result of these studies was the introduction of the boundary temperature of tunneling predominance over activation processes. According to quasi-classical approximation, the probability per unit time of particle tunneling with the mass m for one-dimensional potential barrier of an arbitrary shape (Figure 4.1) has the form:

$$Wo = \nu_0 \exp[-J] \quad J = \frac{2}{\hbar}\int_{a_1}^{a_2} dx\{2m(U(x) - E_0)\}^{1/2}, \tag{4.2a}$$

and for a parabolic potential barrier

$$Wo = \nu_0 \exp\left[-\frac{\pi d}{\hbar}\left(\frac{mE_a}{2}\right)^{1/2}\right], \quad d = a_2 - a_1, \tag{4.2b}$$

where ν_0 and E_0 are the vibration frequency and the energy of a tunneling particle in the initial state; a_1, a_2 are the points of rotation; and d is the barrier width.

By comparing the exponents (4.1a) and (4.2b), we derive for the parabolic potential barrier a temperature criterion below which the tunneling particle transfer proceeds:

$$T_c = \frac{\hbar}{\pi d k_B}\left(\frac{2E_a}{m}\right)^{1/2}. \tag{4.3}$$

With the parameters of characteristic values of chemical reactions $d = 1$ Å and $E_a = 0.5$ eV, we obtain the estimate $T_c \approx 230$ K for H-atom transfer. In the region of higher temperatures, the rate constant has the form (4.1), and at the temperature of the medium approaching the absolute zero it remains constant; i.e., there is a low temperature rate limit. Consequently, in order to obtain the temperature limit in all experimentally studied ranges, it is necessary to explain the intermediate region. This region is characterized by the fact that the mechanism of the particle transfer is of a tunneling character, while the rate constant depends upon temperature (see, for example, References 40 and 41).

The simplest mechanism of the rate-constant temperature dependence is a particle thermal population of the upper energy levels in the initial potential well (Figure 4.1). Under transition from the upper levels, the potential barrier decreases and a probability of transition increases.[1]

As an illustration of experimental data processing on this mechanism, we shall give the results of Reference 42, where for five reactions the H-atom transfer from matrix molecule to methyl radical was studied. Two parameters assumed to be independent were introduced: the height and width of the potential barrier for which the Eckart and Gauss shapes were considered. The best agreement of the experimental data and theory was obtained for the barrier height in the range of 11 to 15 kcal/mol and for the width from 1.7 to 2.3 Å. In this range, the parameters stayed for all five investigated reactions of methyl radical with methanol, ethanol, methylisocyanide, and with two modifications of acetonitril.

At the same time, it is clear that for chemical reactions the width and height of the barrier crossed by the particle are not independent parameters. Besides, a disagreement of the values found in Reference 42 causes some doubt. For instance, if the barrier width lies in the interval 1.7 to 2.3 Å, then, according to simple quantum-chemical calculations, the barrier height corresponding to it must be greater by 3 to 5 times than that obtained in Reference 42.

We shall mention also Reference 43, in which a similar mechanism was investigated using the methods of radiationless transition theory. The reaction was simulated by nonadiabatic radiationless transition between two parabolic terms. The motion on these terms was characterized by frequency ω_0. For the rate constant, an analytical expression was obtained which at low ($T < \hbar\omega_0/4k_B$) temperatures yields the final low-temperature limit. With increasing temperature, this formula corresponds to the activation dependence.

For reactions which were discussed in Reference 42, where H-atom transfer (with a C–H bond dissociation) was considered, the temperature transition to the low-temperature limit ($\hbar\omega_0/4k_B$), according to the results of Reference 43, would be about 1000 K. This is of an order higher than the experimentally observed valences.[42]

Thus, the temperature dependence mechanism which was under discussion in References 1, 42 and 43 does not explain the available experimental data. In the late 1970s, several papers[44-46] were published in which it was shown that it is impossible to explain the experimental data by low-temperature reactions on the basis of the mechanism described in References 1, 42 and 43; however, a new mechanism of the temperature dependence of the solid-phase rate constant was suggested.

Let us determine the range of temperatures where, in the case of a parabolic barrier, a transition to a low-temperature limit takes place. A probability per unit time of tunneling particle transfer from the uppermost vibrational levels would be

$$w_n \approx \exp[-bE_n], \qquad b = \frac{\pi}{\hbar} d \sqrt{\frac{m}{2E_a}}. \tag{4.4}$$

Here, E_n is the height of the potential barrier for a transition from the n-th level. If one assumes that the population of the uppermost vibrational levels obeys the Boltzman distribution (for this the relaxation rate must exceed the tunneling transfer rate), then the ratio of the n-th level population to the ground level population may be written:

$$p_n/p_0 = \exp[-(E_a - E_n)/k_B T]. \tag{4.5}$$

Here p_n, p_0 are the population of the n-th and ground levels of the particle in the potential well. In obtaining the rate constant, it is necessary to take into account the transitions from only the uppermost levels fulfilling the inequality $p_n w_n > p_0 w_0$. By substituting in this inequality expressions (4.2b), (4.4), and (4.5), we obtain the condition of its realization:[45]

$$T \geq \frac{\hbar}{\pi d k_B}\left(\frac{2E_a}{m}\right)^{1/2}. \tag{4.6}$$

By comparing the ratios (4.3) and (4.6), it follows that at temperatures when tunneling prevails thermal occupation of the higher levels may be neglected.

The task of determining the concrete reaction of H-atom transfer from acetonitril molecule to methyl radical was considered in Reference 46. Quantum-chemical calculations have shown that the occupation of the higher vibrational C–H bond levels leads to an increasing tunneling rate constant only at temperatures above 300 K. At the same time, the experiments show that at the change of temperature from 69 to 112 K, the rate constant will increase by the order of two.[46,47]

All this indicates that the increasing probability of tunneling transitions from the upper vibrational levels cannot compensate for a small population value of these levels. This is why the population of upper levels manifests itself only when tunneling does not already play any role and the reaction proceeds in the activation manner.

Another important cause of the rate-constant temperature dependence is the medium reorganization as a result of transition. The first account of a medium reorganization in the task of electron transfer appeared in References 48 and 49, with the intermolecular motions being assumed classical. Later, a theory was developed allowing for the account of quantum properties of the medium and an explanation of experimentally observed charge transfer regularities.[50,51] At low temperatures, the transition rate approaches the finite limit, and at high temperatures the dependence becomes activational. Activation energy depends on medium reorganization energy which determines the rate constant temperature dependence in the condensed phase. This energy is largely in the case of a charge transfer in the polar medium, as well as in reactions accompanied by significant conformation rearrangement of reagents. At the same time, for the reactions of methyl radical with matrix molecules,[41,42,47] which were spoken of above, and where the tunneling transfer mechanism was demonstrated, the rearrangement is small and does not allow the rate constant temperature dependence to be explained. Therefore, if accounting of the reorganization energy for the electron transfer comprehensively explains the process mechanism, then in the case of reactions with participation of a hydrogen atom, a more consistent accounting of medium properties is necessary (see Section 5.3). It follows from all the discussion above that — although the tunneling transition of atomic particles has been studied for a long time and many experimental facts, such as the presence of the low-temperature limit of chemical reactions, were discussed earlier — the theory of this phenomena and a detailed description of the temperature dependence of the transition rate constant were absent and have been developed only recently.

A discussion of the essence of the rate constant temperature dependence mechanism follows. It is known that the probability of a tunneling transition rather strongly depends on barrier parameters overcome by a particle — the probability arises exponentially with decreasing height and width of the potential barrier. Hence, at intermolecular vibrations which cause a change in distance between molecules participating in the reaction, the barrier permeability of the chemical reaction would change and the transition is most probable in the region of the maximum approach of reagents. Thus, the higher the temperature, the larger the amplitude of intermolecular vibrations; the molecules may come closer to each other, the effective potential barrier is less, and, correspondingly, the reaction rate constant is larger. This mechanism of the influence of medium dynamic properties on the regularities of chemical reactions was suggested in Reference 44. It was shown that the intermolecular motions yield the exponential dependence of the rate constant dependence on temperature:

$$k(T) \sim \exp[\alpha T]. \tag{4.7}$$

Here, α is the parameter depending on the medium and reagent properties. It should be noted that the exponent (4.7) is not of the Arrhenius type and is proportional to temperature.

4.2 DESCRIPTIVE METHODS OF LOW-TEMPERATURE SOLID-PHASE REACTIONS

After publication of Reference 44, which for the first time in the theory of solid-phase low-temperature radical reactions demonstrated the importance of intermolecular vibrations, several publications appeared where this mechanism of rate constant temperature dependence was developed.

Before we pass to a brief review of the literature on the given question, we perform a rather simple estimation.[52] It is easy to see that if Equation (4.2) is used, then intermolecular vibrations may influence the parameters $a_1(R)$, $a_2(R)$, and $U(R,x)$. Hence, these parameters and, consequently, the exponent $J(R)$ are the function of the distance between the reagents (R). Figure 4.2 shows how at intermolecular motions the potential barrier, which a particle overcomes as a result of reaction, changes. It is natural that with approaching reagents the barrier is less and its permeability increases.

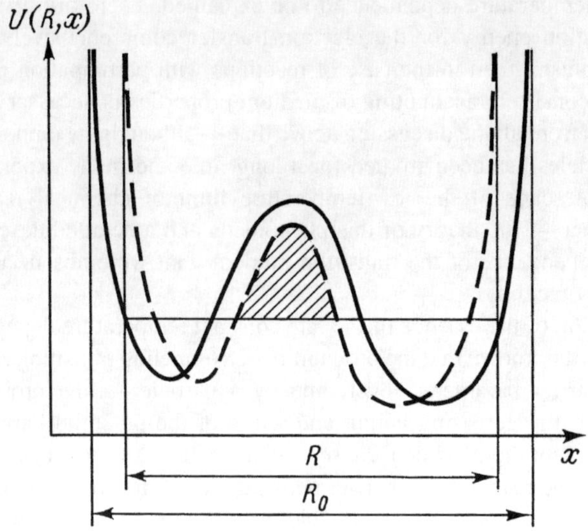

Figure 4.2 Potential energy of tunneling particle. Dotted line corresponds to potential energy approaching the reagents. The potential barrier overcome by the particle has been shaded.

Let us expand the function $J(R)$ in the degree series by reagent displacements from the equilibrium position R_0. In the series, we shall restrict ourselves to a quadratic approximation which for a parabolic barrier is precise:

$$J(R) = J(R_0) + J'(R_0)(R - R_0) + \frac{1}{2} J''(R_0)(R - R_0)^2,$$

$$J'(R_0) = \left.\frac{\partial J(R)}{\partial R}\right|_{R=R_0}, \quad J''(R_0) = \left.\frac{\partial^2 J(R)}{\partial R^2}\right|_{R=R_0}. \tag{4.8}$$

Here, we shall dwell upon the case of classical intramolecular vibrations, and the general case when they can be quantum will be explained below. The probability of reagents approach is proportional to the function $\exp[-\gamma(R - R_0)^2/2k_BT]$ where the γ-force constant corresponds to the relative shift of reagents. Then, the mean probability of transition equals:

$$k(T) \cong \nu_0 \exp\{-J(R_0) + [J'(R_0)]^2 k_BT/2[\gamma + J''(R_0)k_BT]\}. \tag{4.9}$$

It easy to see that the intermolecular vibrations influencing potential barrier parameters lead to the exponential dependence of the rate constant on temperature. Also, it is interesting to note that there is a rather wide part where the exponent linearly depends on temperature. This can be demonstrated if one assumes $J''(R_0) = 0$ in Equation (4.9) (compare with Equation (4.7)).

At higher temperatures, $T > \gamma/J''(R_0)k_B$, the rate constant, assumes the usual Arrhenius type.

It should be noted that by the late 1970s, evaluation was made of the rate constant temperature dependence in the case of quantum intermolecular vibrations.[45,53] The evaluation showed that the mechanism suggested in Reference 44 truly describes transition to the low-temperature limit.

There is an alternative approach to describe low-temperature reactions. The approach is based on the concept of a particle moving over a potential surface of a chemical reaction in the gaseous phase (Figure 4.3). The essence of the given mechanism is contained in the fact that a particle moves not the traditional way along the adiabatic reaction coordinate through the saddlepoint (1 to 4) but instead "cuts off the angle" and passes in the configuration space in a “short way” (2 to 3), tunneling through a sufficiently higher (as compared to the coordinate of corresponding reaction) potential barrier (dotted line in Figure 4.3). The height of the potential barrier is larger but the tunneling length is sufficiently less.

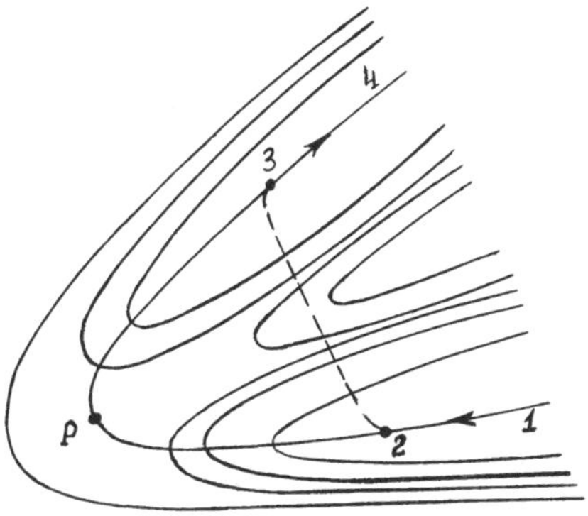

Figure 4.3 Surface of potential energy of gas phase radical reaction RH + R′ → R + HR′; P is the saddle point. (corresponds to the motion along reaction coordinate (1–4); --- corresponds to the underbarrier transition (2–3).

This mechanism was mentioned for the first time in Reference 54. A study was undertaken of the tunneling influence on the transfer rate constant of hydrogen and deuterium atoms at small shifts of a reaction coordinate in the saddle-point region. It was shown that if the de Broglie wavelength of a transferred particle exceeds the characteristic size of the region of the most rapid slope from reagents valley to products valley, then the isolated coordinate of the reaction may not be asserted. Calculations of tunneling corrections were performed for three temperatures: 333, 500, and 1000 K. The tunneling contribution turned out sufficient, naturally decreasing at increasing temperature. Calculations showed that the rate constant has the maximum value at distances between reagents which are somewhat greater than those corresponding to the saddle-point. It is natural that at high temperatures corresponding to the gaseous phase, it was difficult to expect a greater contribution by tunneling.

Some time after Reference 54 was published, several works appeared whereby, using a transient state theory, corrections which the tunneling brings to the atom transfer process were taken into account.[55,56] For instance, Reference 55 considered collinear symmetric H-atom (deuterium) reactions with a hydrogen (deuterium) molecule. The motion along the reaction coordinate was separated from the motion in cross direction where primary attention was paid to accounting for the energy of zero oscillations arising under motions in cross directions. Reference 56 considers development of quantum-mechanical transient-state theory in presenting integrals by trajectories for chemical reactions with atom transfer. Expressions were obtained which can be used for numeric estimations.

The kinetics of tunneling reactions within the scope of a classical S-matrix method were considered in Reference 57. The low-temperature reaction of H-atom transfer between molecules was simulated by a system moving along trajectory 1-2-3-4 (Figure 4.3), taking into account oscillations in the valleys of reagents and products. The under-barrier transfer was described by imaginary time which was assumed small compared to the oscillating period of heavy particles. Numeric calculations of temperature dependence and the isotope effect of linear H-atom transfer reactions for characteristic surfaces of potential energy were conducted.

Reference 58 also should be mentioned. The authors studied symmetric collinear reactions of H-atom transfer between two heavy particles with fast H-atom motion proceeded on the background of slower motion of heavy particles. The calculations used polar coordinates. Along the radical coordinate, the translational motion of heavy particles proceeded and the H-atom motion was described by the angle coordinate. A more complicated case of asymmetric reaction was considered in the work.[59]

All these calculations, as mentioned above, were carried out for a potential energy surface of the gaseous phase reaction corresponding to Figure 4.3. At the same time, in the range of temperatures corresponding to the achievement of a low-temperature plateau, a substance usually becomes solid and the motion of reagents is limited. The surface of potential energy for solid-phase reactions differs from the surface shown in Figure 4.3. It is clear that in the initial position the reagents are at the final distance from each other and after its completion the products do not scatter.

A characteristic view of the potential energy surface corresponding to solid-phase reaction is shown in Figure 4.4. Reagents vibrate around the initial state (1) and come close (1–2), and in the most suitable configuration a transition occurs (2–3). When the products occupy (3–4), a new position of equilibrium (4) is reached. It is this mechanism to which expressions for the rate constant (4.7) and (4.9) describing an atom transfer in the solid phase correspond. These expressions hold true in the case when intermolecular vibrations may be considered classical. In fact, the most interesting is a temperature range when intermolecular vibrations become quantum. That is where the connection of lower temperature limit lies.

Such a consideration of low-temperature radical solid-phase reactions was carried out in References 60 to 62. Within the framework of a modified theory of radiationless transitions, expressions were provided for atomic transfer rate constants over a wide range of temperatures, for helium to room air. Different aspects of this theory and results of its application will be considered in the following sections. Here, we shall dwell briefly on conclusions.

Three main temperature ranges may be separated (see Figure 4.5).[9,63] The first is a low-temperature interval (I) when both a particle transfer and a motion of molecular environment are quantum; moreover, temperature dependence of the rate constant is determined by the type of intermolecular vibrations: in the case of a local mode and the Einstein descriptive model of molecular motions, the dependence

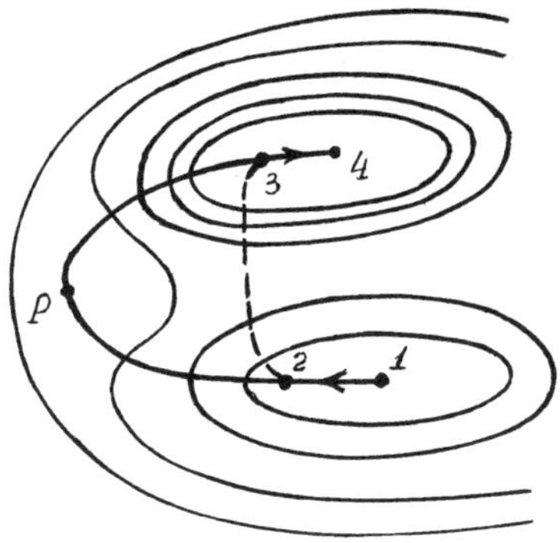

Figure 4.4 Surface of potential energy of solid phase radical reaction. The reagents oscillate near the initial (1) and final (4) equilibrium position.

looks like a low-temperature plateau, and for the ideal crystal lattice (the Debye model) the logarithm of the rate constant is proportional to the fourth degree of temperature. The next part (II) corresponds to classical intermolecular motions and quantum transfer of a particle. The logarithm of the rate constant in this range is proportional to the temperature. Finally, in the range of sufficiently high temperatures (III), both intermolecular vibrations and particle transfer are classical. The rate constant dependence on temperature is of the Arrhenius type.

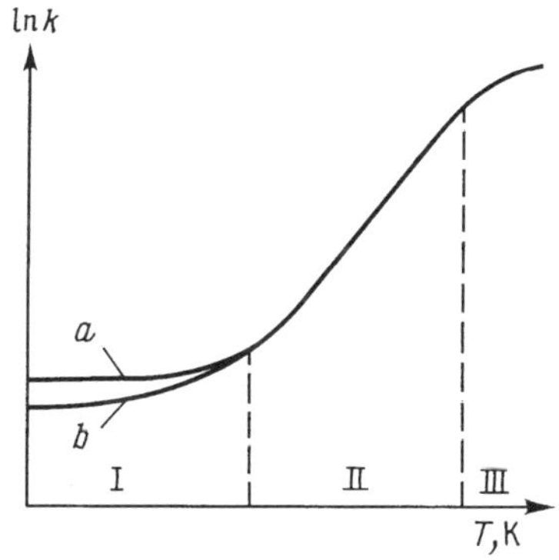

Figure 4.5 Temperature dependence of rate constant of solid phase chemical reaction. I = low temperature part of the curve (*a* — local mode, *b* — ideal crystal). II = linear section of the curve (see Equation (4.7)). III = activation-like part of the curve (see Equation (4.9)).

In this section we shall move on from discussion about the results of our works on the theory of solid-phase reactions (Sections 5.2 and 5.3 will explain this question) and will consider theoretical papers of other authors who, in our opinion, have made the most contributions to exploring this problem.

The first work which supported the above-mentioned mechanism of rate constant temperature dependence was Reference 64. Attention was devoted primarily to the problem of nonadiabativity of atom transfer in chemical reactions. Calculations performed within the framework of the general radiationless transitions theory are conducted under the assumption that tunneling is instant. At the same time, a theory constructed taking into account the transfer time (τ) shows the importance of the value $\omega\tau$ (ω is the frequency of intermolecular vibrations). If this product is of the order or more of the unit, then the vibrations of environmental molecules which do not change distances between reagents are also sufficient.[64] Simple estimations show that accounting for nonadiabativity effects is necessary only for the transfer of heavy atoms with a mass greater than H-atoms, and the influence of intermolecular vibrations on potential barrier parameters should be taken into account for hydrogen atoms, too.

At approximately the same time a paper was published which also considered thermal vibrations of reagents altering the potential barrier width of the reaction.[65] From extension (4.8), the first two terms were taken into account, which, after averaging the classical transition probability, leads to a linear dependence of the exponent on temperature — which corresponds to expression (4.7). A comparison of the experimental data on low-temperature reactions of acryl radicals[66,67] isomerization and formation of hexahydrocarbazol[68] with this ratio was carried out and a rather good agreement was demonstrated. Other mechanisms of temperature dependence were discussed as well, but they could not compete with the thermal barrier fluctuation method on accuracy within reasonable parameter values of the experimental data of reproduction.

In contrast to References 64 and 65, which confined themselves to considering the first two terms from formula (4.8), References 69 and 62 (see also Reference 6) used this ratio as a whole. According to the Fermi golden rule which was used earlier[6,60-62] to describe atomic tunneling, a standard formula for the H-atom transfer rate constant between two heavy particles was written. It was reported that relative displacements of reagents with defined frequencies take place. Unfortunately, in the rate constant expression the authors failed to carry out the summation analytically, as was done in References 6 and 62. The rate constant was analyzed numerically. The temperature dependence of the rate constant was obtained in ranges from a low temperature plateau to activation-like law for concrete examples of radical reactions of H-atom abstraction from methanol molecules by a methyl radical and the processes of intramolecular H-atom transfer in phynyl radicals.[67] A good agreement of the theoretical curve with experimental data is observed.

The experimental data processing technique, developed in Reference 69, was utilized in two later works[70,71] to study H-atom transfer in the fluorene-acridine system.[72,73] The problem of adjusting theoretical results to the experimental will be discussed in Section 5.5, therefore, we shall only note that in References 70 and 71, expression as used for the H-atom transfer rate constant when the oscillation of intermolecular distance proceeds with a defined frequency. At the same time, tunneling in an ideal crystal[72,73] in the discussed experiments, was studied where, as it is known, dispersion ratios for the frequency exist.

In Reference 74, a comparison was performed of the exponent describing the temperature dependence of H-atom transfer with experimental data which had been obtained earlier.[75] Photochemical formation of radical reactions was studied in molecular crystals of fluorene with small additions of acridine. The rate constant (k) of H-atom transfer was measured in the range of 60 to 300 K.[75] It was found, that as predicted in Reference 44, experimental data in coordinates *lnk-T* fitted well in the direct line (compare with Equation (4.7)). Attempts to describe experimental data by a sum of several Arrhenius exponents did not yield positive results. The authors were only able to conclude the tunneling mechanism of the reaction.

To study tunneling phenomena, a matrix density[76] apparatus is used. Tunneling transitions in asymmetric double-well potential corresponding to the state $2B^1\Sigma^+$ in the molecule HCl were considered. The use of matrix density apparatus allowed for accounting of the studied system's interaction with the environment. It is clear that such an interaction[76] may cause potential fluctuations leading to changes in the barrier permeability which were considered in References 6, 62, and 69.

It was assumed in the above-mentioned works that a change of distances between potential wells happens as a result of intermolecular vibrations. This is true only when the particle transfer process from one well to another is accomplished in the solid phase at low temperatures with increasing temperature. Especially in the solvent phase it is necessary to take into account a diffusion character of reagent motion. It is this consideration that was undertaken in Reference 77. A tunneling transfer of proton between molecules was considered, the molecules being simulated by two similar potential wells. It was assumed that the wells are spherical and rectangular, and their approach is described in the framework of the classical diffusion relative motion of molecules. In the work, a level splitting is determined and the rate constant of resonance proton transition is found. Hence, the diffusion to tunneling process was shown. As a result of the study, the authors verified that decreasing temperature leads to prevalence of the tunneling mechanism over ordinary activation transfer. The isotope effect in the substitution of hydrogen atom by deuterium was studied. It turned out equal to the value 10^7, which noticeably exceeds the experimentally obtained value.

The tunneling transfer of hydrogen atoms and protons in solutions was considered in Reference 78. Apart from the oscillations of potential barrier parameters, medium reorganization and energy defect of fast subsystems were taken into account. Only the linear term from expansion (4.8) was accounted for. Expressions for the tunneling rate constant and different partial cases were obtained. It should be noted that the main formula describing the rate constant dependence on the above-listed factors was obtained earlier in Reference 62 by another method. A very close approach with the same operator of interaction was used in Reference 79. The tunneling system was represented by pseudospin ($S = 1/2$) and was bound with an environment modelled by harmonic oscillators. These oscillators play an active part in the transfer process. A case of high temperatures and transition to activation dependence was considered. However, for a corrected transition to activation dependence, accounting for the quadratic term from expression (4.8) is necessary and this was not done in Reference 79.

Later, a molecular-dynamic modelling of a proton transfer symmetric reaction in solutions was carried out.[80] Potential wells, between which a proton transfer takes place, participate in oscillations modelling a chemical reaction barrier; however, they are influenced by a transfer process stimulating environmental fluctuations, which lead to a disruption of the potential symmetry. The important role of proton interaction with the environment and inter- and intramolecular vibrations was demonstrated.

The author in Reference 81, after Reference 62 (see also Reference 6), used the quadratic form for the exponent (4.8). In the framework of a modified theory of radiationless transitions, the rate constant of H-atom transfer in dimethylglioxime was calculated.[6] The Debye and Einstein models of a solid are discussed. A conclusion[81] is made that a unimode medium model (the Einstein model), allowing a description of potential barrier parameter oscillations, may yield the same effect as the Debye model, i.e., it is asserted that the rate constant temperature dependence will not change if instead of the dispersion spectrum of crystalline lattice oscillations a unimode model with effective mass is used.

Quite recently, a study of the temperature dependence of tunneling chemical reactions was carried out.[82] Symmetry potential barrier was calculated for H-atom transfer reactions from methane molecule to methyl radical. This barrier was modelled by the symmetric Eckart function, for which an analytical expression is available describing the tunneling permeability of the barrier. Further account was taken of tunneling probability dependence on atomic energy making a transition. Both the discrete energy of C–H oscillations and the energy of low-frequency vibrations of the reactionary were studied. To obtain the rate constant, the transition was averaged by all types of oscillations. This work is very close to the work of Reference 42 as is shown above; see Equations (4.3) and (4.6), as well as explanations for them. Such a mechanism cannot explain the rate constant temperature dependence of tunneling reactions. The authors point out that the formation of temperature dependence of the rate constant must be determined by oscillations of the reaction system,[82] however, the calculation method used cannot give an adequate description of experimental data.

It also should be said that atomic particle tunneling has been studied for over a decade within the scope of Kramer's problem when the dissipation of energy in the transfer process is taken into account. Development of this approach started from published works which considered a task of dissipative tunneling at $T = 0$.[83-85] The interesting qualitative result of these works is in the fact that the dissipation of energy exponentially decreases the rate constant of tunneling. This leads to decreasing the limiting temperature of tunneling predominance over the Arrhenius.

Later, much attention was paid to different aspects of this task (see, for example, Reference 8, 11, 86–89). However, as mentioned in Reference 8, the majority of works were performed in the Condon approximation without taking into account the matrix element transfer dependence on photon coordinates, i.e., for the one-dimensional potential barrier. It is necessary to point out that even in those cases when such an accounting has been carried out,[86] this dependence arises only because of the interaction of the tunneling particle with the vibrational reservoir (otherwise, account is taken of the interaction of intra- and intermolecular modes rather than the decreasing potential barrier as a result of intramolecular vibrations). The correctness of this statement follows from the fact that no difference in the cited works is made between the inter- and intramolecular transfer of particles. Nevertheless, it is obvious that in the case of atom transition from one molecule to another, the rate constant temperature dependence is determined not by the interaction between different modes but by the influence of changes in reagents arrangement on the potential barrier of chemical reaction.

Also, the role of interaction between modes is discussed in Reference 90, which considered radiationless transitions in the non-Condon approximation, in addition to the expansion of the matrix transfer element to a degree series by nuclei displacement. The question of the rate constant temperature dependence mechanism in chemical reactions was discussed in Reference 91, which investigated the role of interaction between inter- and intramolecular vibrations. The exponential type of operator interaction between modes was described.

Some works have been devoted to studying the intramolecular transfer of proton, for example, a numeric modeling of the reaction of a two-proton transfer in the dimere of formic acid.[92] This process was studied earlier assuming that the geometry and electron structure of molecules, in which transition occurs, have equilibrium. Using a designed *ab initio* three-dimension potential surface, the reaction path of proton transfer with minimum energy has been found. That is, it has been found that in the reaction process a molecule's geometry may alter. It is shown that protons move along a tunneling short path rather than the reaction coordinate. Parameters of the potential barrier crossed by a quasi-particle are calculated (synchronic motion of two protons). These studies were continued in Reference 93. Along with the reaction of simultaneous transfer of two protons, proton tunneling in malonaldehyde was studied which was accompanied by transfer of a double bond. The problem is in the search for "the maximum probability way". The basis of this approach is a very precise *ab initio* calculation of the ground state wave function.

Consideration is confined to symmetric exchange reactions, however, this method can be extended to other types of reactions. The same task of proton intramolecular transfer was solved by using Feinman integrals over trajectories.[94] By means of this method, summations were performed in the expression for the rate constant which is based on the known Fermi golden rule. Account is taken of both the polar fluctuations of the medium and the barrier preparation under motions by other coordinates connected with the reaction mode.

The numeric modeling of proton transfer in the quasi-symmetric fragment O–H–O was carried out in Reference 95. A surface of the potential energy for this reaction was constructed. The proton transfer (a fast subsystem) was modelled by oscillations of the potential wells (a slow subsystem). The possibility of proton tunneling from uppermost vibrational levels in a potential well is considered. Transfer rate constants from uppermost vibrational levels of both inter- and intramolecular subsystems were calculated.

Among works carried out recently we shall note Reference 96, which reports quantum-chemical calculations of a potential barrier for proton transfer in malonaldehyde. A proton transfer when the molecule stays both in ground and excited states was considered. Reference 97 is of interest because it theoretically investigated the intermolecular proton transfer in dimethyltetralon. The barrier preparation as a result of intramolecular motions and medium rearrangement was taken into account. Earlier, a simultaneous accounting of these phenomena was performed.[6,62,98,99] Questions of tunneling splitting of levels were studied in Reference 100. A multi-dimensional tunneling in a quasi-classical approximation was explained. The authors of Reference 101 have developed a method to reduce the transient state of the reaction system which was used to discuss the regularities of the reaction

$$Cl = HCl \rightarrow ClH + Cl,$$

investigated experimentally earlier.[102] Reference 103 joins the above-discussed problems with its discussion of the possibility of particle motion not by reaction coordinate. The question is about diffusion of a light admixture in metals when a particle moves in the external field without forming and breaking down chemical bonds and the relief is formed by outer atoms of the metallic lattice. We shall note that the consistent quantum theory of light admixture diffusion in the external field of the crystalline lattice has been developed.[104]

4.3 LOW-TEMPERATURE RADICAL REACTIONS EXPERIMENTAL DATA

In this section, radiation-chemical and photo-induced processes where the rate constant, low-temperature limit was observed shall be considered. Briefly the essence of the phenomena is as follows. In the range of sufficiently high temperatures the rate constant dependence, as is known, is of the Arrhenius type. With decreasing temperature, a decrease of the effective activation energy occurs and as a result the rate constant becomes equal to the finite value independent of temperature. Thus, the finite rate limit of chemical reaction is achieved.

It should be said that small deviations from the Arrhenius law in gaseous and solvent reactions which demonstrate the presence of a tunneling correction have been observed.[105] Moreover, the valid low-temperature plateau, as expected, was observed only in solid-phase reactions and discovered in studying an electron transfer. A process of citochrome C oxidation by chlorophyll was investigated, and it was found that the characteristic time of electron tunneling does not depend on temperature in the range of 4.2 K to $T_c \approx$ 120 to 130 K.[106] In a study reported in Reference 107 electron transfer from anion-radical O^- from traps was established. The low-temperature plateau for this process started in the range of 100 K.

About the same time that Reference 106 appeared, the first indications of the existence of tunneling reactions with transfer of heavy particles were reported. Also, radiation polymerization reactions of acrylonitrile and tetrafluoroethylene at helium temperatures were studied.[108,109]

However, these experiments did not allow the γ-radiation role to be determined. To be more precise, they did not answer the question of whether the chemical reaction proceeded after radiation creation of the chain spontaneously or was the effect of radiation sufficient in the course of the whole process of the chain growth. Later, this question became the subject of investigations.[110]

In the following years the kinetics of a large number of solid-phase reactions were investigated. A detailed, sufficiently complete description of experimental data on low-temperature reactions is given in Reference 10. Some typical temperature dependences are given in Figure 4.6. These reactions can be characterized by several parameters: low-temperature limit value, T_c, and the temperature value at which an increasing rate constant and effective activation energy begin. It is easy to see that several groups of reactions may be isolated for which these parameters are sufficiently different. First, the low-temperature limit value in the case of heavy atoms transfer (Cl and Br) is of several orders greater that a similar value for intramolecular H-atom transfer. Certainly it is more natural if the tunneling of a heavy atom proceeds slower than the H-atoms transfer, but a sufficient difference of particle masses can be compensated for by the difference in parameters of potential barriers. It should be noted that in this case the role of above-barrier transitions become rather essential.[111,112] A second important difference of the rate constants of these reactions consists of the values of temperature T_c for such processes as polymerization of formaldehyde ($T_c \approx$ 10 to 20 K), intermolecular H-atom transfer ($T_c \approx$ 50 K), and intramolecular H-atom transfer ($T_c \geq$ 100 K).

Further, the indicated processes will be discussed and mechanisms of their flow suggested. We shall note only the following. It is assumed that any reaction with a low-temperature limit proceeds in the region of a low-temperature plateau according to a tunneling mechanism.[10] And, if the reaction mechanism is of a tunneling character, then a low-temperature plateau is observed. However, as shown below (see Sections 5.2 and 5.3), there is a sufficiently large area where the rate constant very strongly depends on temperature, and the transfer mechanism is tunneling. On the other hand, the presence of a low-temperature limit does not always guarantee the tunneling mechanism of chemical reaction (Section 6.2.3).

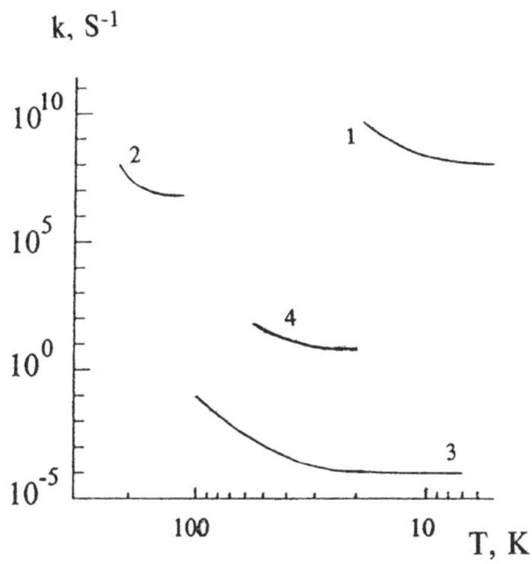

Figure 4.6 Typical temperature dependences of different kinds of solid-phase chemical reactions in which low temperature limit was observed: (1) low-temperature reactions with a limiting physical stage; (2) tunneling of H-atom in monomolecular reactions; (3) low-temperature limit in the reactions with intermolecular H-atom transfer; (4) low-temperature reactions with transfer of heavy particles.

4.3.1 Low-Temperature Reactions with a Limiting Physical Stage

The first attempts to study the process of γ-induced polymerization at $T = 4.2$ K led to the discovery of the low-temperature rate limit of chemical reactions.[108,109] The radiation-induced polymerization of formaldehyde over a wide range of temperatures (4.2 to 140 K) was studied. A low-temperature plateau is observed, and in the range of 10 to 12 K an increasing rate constant begins and the temperature dependence becomes activation-like. The studies were conducted by means of a sensitive calorimeter, whose relaxation time was of the order of 1 second.[113]

The process of the polymer chain growth includes a charge transfer toward the end of the chain as well as the motion of a formaldehyde molecule. The charge transfer cannot be a limiting stage, since in this case an increasing rate constant must start at sufficiently higher temperatures. For this reason, the specific temperature dependence features of the reaction considered should be grouped together, namely, with the mobility of a formaldehyde molecule. It has been pointed out that the polymerization temperature dependence of formaldehyde can be determined by orientational motions and the adherence of a fragment to the chain proceeds without overcoming the potential barrier.[114] In fact, defreezing orientational oscillations of small molecules takes place in the range of the polymerization rate constant.[115] Thus, there is an agreement between the temperature dependence of the chemical process and orientational motion.

A postpolymerization of formaldehyde was observed as well.[116] This polymerization was initiated by γ-radiation which was switched off by magnetic dumper. The experiments were carried out at a temperature of 5 K. The polymerization of formaldehyde after switching-off proceeded for 20 min. Thus, it was demonstrated that the reaction of the CH_2O fragment connection at super low temperatures may proceed spontaneously without direct radiation exposure. Discussion of different mechanisms of chain growth under polymerization of formaldehyde is given in Reference 117.

The next example of low-temperature solid-phase reactions is the rebinding of ligand bonds with the iron of the gem-group mioglobine.[118-120] Under exposure to a 1-μsec laser pulse, this bond ruptured, after which its rebinding occurred and was registered spectrographically. The process whose kinetics is studied consists, evidently, in the yield of the iron atom from the gem-group plane and its following return. Izon atom displacement by photoabstraction of oxygen molecules or carbon oxide (ligands) is initiated. The recovery of the ruptured bonds is an exothermic process which includes the stage of ligand displacement and, possibly, its rotation. The plateau in the rate constant temperature dependence observed below 10 K can be associated with such a rotation.

These investigations were carried out over a wide range of temperatures, from 2 to 320 K. It is interesting that below temperature 200 K, polychronic kinetics were observed typical for low-temperature reactions in glasses.[121] Starting from the tunneling mechanism of the reaction, a selection was undertaken of the parameter value spread of the potential barrier crossed by a particle.[119] At the same time, all these calculations have a rather formal character. Even if a model suggested holds true, it is not clear what mass is tunneling.[118,119] Of course, the length of ligand tunneling can exceed sufficiently the distance over which the iron atom moves. But the process of transition of the iron atom can be limiting. Construction of different models describing the kinetics of rebinding ligand bonds continues.[122,123]

The kinetics of such processes under substitution of CO ligands by different isotopes have been investigated.[120] The results of this work, in the opinion of the authors, directly pointed to tunneling, in a rather complicated potential consisting of several potential wells separated by barriers.[120,122]

The decomposition of biradical 1,3-cyclopentadyil in the cyclohexane matrix was studied. 1,3-cyclopentadyil was obtained at photolysis of molecules 2,3-diazabicyclo-2,2,1-hexane-2. The photon source was switched off and the attenuation of an ESR signal was observed. The minimum temperature at which this spectrum was observed was 1.3 K.

The kinetics of the studied process turned out to be polychronic; therefore, the rate constant values were determined by the initial part of the kinetic curve. The range of temperature at which the rate constant can be considered accurately measured makes up some tenth of degrees — from 5.5 to 40 K. The low-temperature plateau started below 20 K; at higher temperatures the rate constant dependence is the Arrhenius one.

Interpretation of the experimental data on the basis of the tunneling mechanism was performed. The values of the barrier parameters turned out to be fully reasonable (V = 2.3 kcal/mol, d = 0.64 · 10^{-8} cm, and m = 14 unit H-atom mass, which corresponds to the CH_2 group; the barrier was assumed parabolic). However, these parameters can be considered reasonable only individually. It is completely obvious that the height corresponding to the given barrier width must be much larger.

It should be taken into account that in order to convert the biradical, apart from the bond formation, a small change in the configuration of its skeleton is necessary. It is this that can cause the temperature dependence observed.

In the inert matrix of nitrogen the reaction between molecules of ozone and nitrogen oxide was investigated.[125] The complex ($NO \cdot O_3$) was formed in the matrix which under the effect of infrared radiation decayed into oxygen and nitrogen dioxide. The activation energy of this reaction measured in the range of 11 to 20 K was extremely small, a little more than 0.1 kcal/mol.

The work discussed the reaction tunneling mechanism and on it based the estimation of the parameter values of the potential barrier. The importance of orientational motions was noticed where the activation energy value of these motions corresponds to the measured one.[115,126] After the rotation of molecules, the reaction may proceed with zero activation energy.

4.3.2 Tunneling of a Hydrogen Atom in Monomolecular Reactions

The low-temperature limit in reactions with intramolecular H-atom transfer was first observed by the authors of References 66 and 67. The ESR method was used to study sterically hindered isomerization of aryl radicals in hydrocarbon solvents. The temperature range where the investigations were carried out was 28 to 240 K. The low-temperature limit was achieved in the range of 100 K.

In the approximation of one-dimensional tunneling by the reaction coordinate, the calculation of the transfer rate constant for different shape barriers was carried out. The agreement was not achieved for any shape of a potential barrier.

These reactions also were studied in deuterated compounds in the range of temperatures 123 to 233 K, with the low-temperature limit not being found. There was observed only some deviation from the Arrhenius law. The isotope effect of this reaction at temperature 123 K is 10^4.

The next monomolecular process of H-atom transfer at low temperatures has been studied.[68,127] A sigmathropic shift in the light-excited triplet state of the derivative diphenylamine was studied. The reaction studied proceeds in two steps: excitation of diphenylamine to a triplet state (with bonds between rings being formed) and transfer of hydrogen atoms. In the second step at temperatures lower than 100 K the low-temperature limit was observed. The reaction was investigated in the range of 80 to 290 K. The isotope effect, 5 · 10^3 at T = 100 K, was discovered to decrease with increasing temperature. In the approximation of the one-dimensional Ekkart barrier, the calculation of the rate constant temperature dependence was carried out. However, the calculation results agree with the experiment very poorly.

A study of the tunneling atom transfer between the atoms of carbon and oxygen in the molecule of methylacetophenon has been carried out.[128] Mixtures of this substance with organic compounds were exposed to light after which keto-enol transitions were observed. The measurements were conducted in the temperature range 92 to 313 K. Below 150 K the Arrhenius law for this curve was not realized. Also, an isotope effect was discovered to be equal to 180 ± 20 in the range of temperatures 118 to 140 K and decreased up to 3 at $T = 298$ K.

These experimental data are compared with the calculation of the tunneling transfer rate constant for the Eckart symmetric potential barrier. It cannot be asserted that agreement of the calculation and experiment is achieved at reasonable values of the barrier parameters.

A slight difference of the rate constant temperature dependence from the Arrhenius law was observed in the process of tautomerization of the excited 7-azaindole at temperatures of 170 to 280 K.[129] The magnitude of the isotope effect increases from 3.2 at 272 K to 14 at 172 K.

The rupture of the O–H bond with the creation of an N–H bond was studied in photoexcitation of the 2-(2′-hydroxyphenyl)benzoxazole (HBO) molecule in organic solutions.[130] The temperature range for the investigation was 55 to 200 K. A noticeable deviation from the Arrhenius law arose in the range 110 K, then the dependence leveled out ($T < 70$ K). The isotope effect increased from the value 30 at 200 K up to 300 at 100 K. Similar dependences were observed in 5,8-dimethyl-1-tetralone (DMT systems) at transition from keton form to enol.

The process of inversion of oxiranil and dioxolanil radicals has been studied.[131,132] The shape of the ESR spectrum of these radicals depends on the frequency of H-atom transitions between double equivalent states. At the same time, the transfer frequency (see Sections 5.1 and 5.1.2) naturally is determined by the potential barrier which separates these states. The transfer frequency studied corresponds to splitting in the ESR-spectrum which is observed experimentally.

Measurements of the frequency of the H-atom oscillations in the oxyranil radical were conducted between 105 and 223 K.[131] The low-temperature plateau started at 140 K. The calculations of the tunneling temperature dependence with use of a nondimensional potential barrier yielded fairly good agreement with the experiment.

A rather close process of H-atom transfer in a dioxolanil radical was studied in the temperature range 92 to 142 K. The low-temperature limit of the rate constant was achieved at lower temperatures ($T_c \cong 100$ K) as compared with the inversion reaction of the oxyranil radical. This process is more complicated than the tunneling oscillations in the oxiranil radical. In the given case the potential barrier separating the equilibrium positions of the hydrogen atom is modelled by low-frequency outside-plain vibrations (the vibrational frequency is equal to 140 cm^{-1}). It is shown that only by taking into account that the space of the potential barrier is two-dimensional, it is possible to obtain satisfactory agreement of theory and experiment. A comparison of the potential barrier found on the basis of experimental data and the barrier has been calculated *ab initio*.

Note that in experiments the frequency of probability oscillations finding a particle in some potential well is noticeably higher than the rate constants of the relaxation processes in the initial and final states.[131,132] This provides a possibility to observe splitting in the ESR spectrum. The oscillating frequency is proportional to the matrix element of transition (see Sections 5.1.1 and 5.1.2) rather that its square as in the case of irreversible processes. The calculations were designed to obtain transfer rate constants, and a coherence possibility of the tunneling process was not taken into account.[131,132]

Some more complicated phenomena were observed in HBO molecules and *m*-MeHBO excited-to-triplet states. These molecules are suitable to observe the process of intramolecular transfer since a hydrogen atom has two closely located equilibrium positions. In one of them, the hydrogen atom bonds with the oxygen atom (the enol form — *E*). This is the ground state of the molecule. The second energetically higher state corresponds to the keton form — *K*. The oxygen atom is bonded with the nitrogen atom.

Absorbing the light, the molecule transits from the ground state ^1E to the higher singlet level $^1E^*$. The molecule performs a transition to the lower closely located singlet state of the ketone form $^1E^*$. After that as a result of intercombination conversion (see Section 1.3.2), excitation of the ketone triplet state $^3E^*$ takes place. It was shown that between two isoenergetic triplet states ($^3K^*$ and $^3E^*$) during several nanoseconds equilibrium is established.[130] These established states slowly disintegrate as a result of phosphorescence and radiationless transitions. A study of phosphorescence of this system allowed the rate constant temperature dependence of H-atom transfer between states $^3K^*$ and $^3E^*$ to be found.

Similar processes may take place in the molecule *m*-MeHBO; however, there is a slight difference. The triplet level of the enol form lies below the triplet state of the ketone and, consequently, at low temperatures transitions $^3K^* \rightarrow {}^3E^*$ will prevail. It is necessary to take into account that the distance between the levels is sufficiently large, 330 cm^{-1}. This value noticeably exceeds the energy which corresponds the temperature of the medium. The regularities of the phosphorescence of compounds HBO and *m*-MeHBO have been studied in detail.[134]

Measurements of the rate constant temperature of the tunneling transfer (H-atoms) were conducted in the range between 35 and 250 K. In both compounds the rate constant ceased to depend on temperature in the interval 80 to 100 K. Also, the temperature dependence of the isotope effect of these reactions was measured. Its value at reaching the low-temperature limit somewhat exceeded the value 10^3. The temperature dependence of the rate constant was investigated in different solvents in which the friction factors noticeably differ.[133] For instance, in trimethylpentane the friction factor is of an order larger than the friction factor in trimethylpentane — isopentane. At the same time, the results of measurements for these compounds do not differ substantially. A conclusion can be made that in the above-mentioned range of temperatures friction does not affect the reaction rate constant. With further decreasing temperature, and this is noted in Reference 133, friction may manifest itself.

Thus, it follows that the probability of H-atom transfer in the molecule HBO depends also on the value of constant longitudinal and transversal relaxation of the states $^3K^*$ and $^3E^*$. These constants also are temperature functions. That is, at super low temperatures the reasons for a rising rate-constant temperature dependence may be different.

To transfer hydrogen atoms, the kinetics of reaction in HBO is monoexponential only in the solvent phase. In solid solutions the HBO kinetics of transfer is polichronic, which is usually explained by the environmental influence on the potential barrier of the reaction studied. As to the chemical process in *m*-MeHBO, here the kinetics is monoexponential. At the same time, the most significant difference of

HBO from *m*-MeHBO is contained in the value of the resonance disturbance between levels $^3K^*$ and $^3E^*$. Consequently, the absence of the nonexponentiality of reaction kinetics in *m*-MeHBO might be explained by the smallness of the relative change of resonance disturbance under environmental effect.

The presence of several mechanisms of chemical reactions may be one more reason for the nonexponentiality of the solid-phase reaction. For instance, under ultraviolet irradiation N-ethyl-2,6-dimethyldiphenylamine is cycled with the formation of zwitter ion dihydrocarbazol.[135] The latter then is isomerized in stable dihydrocarbazol as a result of H-atom transfer. It is this stage that is limiting.

The range of temperatures where the reaction was studied is approximately 68 to 293 K. Up to temperature 240 K the reaction remains monoexponential. At further decreasing temperature, the process becomes nonexponential and is described by the combination of monomolecular and bimolecular reactions. The low-temperature limit for the monomolecular reaction of hydrogen atom transfer is achieved in the range 160 K, for the bimolecular, at 110 K.

The process was studied in different media where the viscosity changed on the order of four. The kinetics of the monomolecular transfer does not depend on viscosity and for the bimolecular transfer such dependence is not observed. It is clear that viscosity starts influencing the kinetics only when the kinetic mode of reaction transits to the diffusional. This is what happens in the range of temperatures 80 to 100 K.

4.3.3 The Low-Temperature Limit in the Reactions with Intermolecular H-Atom Transfer

Let us move on to discussion of chemical reactions with tunneling intermolecular transfer of hydrogen atoms. The low-temperature limit in such processes was first observed in Reference 137. In the range of temperatures 15 to 100 K, the radical reaction was investigated:

$$CH_3OD + CH_3 \rightarrow CH_2OD + CH_4. \qquad (4.10)$$

The methyl radical was formed from molecule CH_3Cl at the dissociative capture of the electron which was produced as a result of light absorption by the readily ionized admixture (TMPD). The radicals were registered using the ESR method.

It should be noted that the nonexponential kinetics observed were explained by the matrix nonuniformity. In order to calculate the rate constant, the initial part of the kinetic curve was used under the assumption that the reaction kinetics may be described by the first order law. The rate constant temperature dependence of reaction (4.10) was obtained; the low-temperature plateau was achieved in the range 50 K. At higher temperatures, the rate constant increases rather rapidly.

Experimental data were processed by using a procedure based on the tunneling mechanism of the reaction.[138] Good agreement of theory with experimental data was obtained only in the range of temperatures 60 to 85 K. At lower temperatures, the calculation results for one-dimensional tunneling differ significantly from the experimental; in particular, the activation energy value was very low.

A study of the low-temperature methyl radical reactions in alcohol matrices was continued.[42] The reaction of H-atom abstraction by a methyl radical from the ethyl alcohol molecule was investigated; the procedure of measuring and preparing mixtures coincided with the procedure described in the work.[137] The range of temperatures at which the measurements were conducted was 13 to 99.2 K. The low-temperature plateau started approximately at the same temperatures as in the methanol matrix. Primary attention in this paper was paid to discussing a procedure for processing experimental data (see Section 4.1); however, a good agreement between the experiment and the theory at reasonable parameter meanings has not been reached.

In addition to this work, another published paper reported the temperature dependence of H-atom transition rate constants in irradiated dimethylglyoxime.[139] As a result of γ-radiation there were radical pairs of two kinds in the sample. The registration was carried out by the ESR method. As shown earlier,[40] by the same means of radicals generation and at temperatures over 77 K, transition of one kind of radical pairs into another took place. The temperature range was extended (4.2 to 140 K) and it was found that below 50 K a low-temperature plateau begins.[139]

A comparison of the theory with the experimental data was carried out. The tunneling constant was calculated for the Eckart barrier at 0.8 Å wide and 15 kcal/mol high. Above 50 K a fairly good agreement between the theory and the experiment was observed. However, at 4.2 K the theoretical meaning of the constant exceeded that of the experimental.

The most detailed study of low-temperature reaction regularities was reported in References 72 and 73. These works are the continuation of studies carried out at higher temperatures.[74,75] The reaction of H-atom transfer in the crystal of fluorene with an admixture of acride was studied in a range of temperatures 33 to 300 K. The acridine molecule was excited by optical radiation to a higher singlet state, then by means of intercombination conversion a population of the lower triplet level took place. This state could decay due to double processes — triplet-singlet transition to the ground state and the H-atom transfer from the fluorene molecule to the excited acridine molecule with formation of a radical pair.

It was found that the rate constant of the intramolecular transition depends very weakly on temperature.[75,140] For instance, at room temperature the value of this constant is only two times more than its low-temperature limit. The total value of the rate constant low-temperature limit of the triplet state decay is equal to 13.88 sec^{-1}, and at low temperatures the tunneling rate constant must meet the inequality:

$$k(0) << 1\ s^{-1}. \tag{4.11}$$

A very important feature of this work consists of the fact that the H-atom transfer does not distract the fluorene matrix. Meanwhile in the case of tunneling in the crystal of dimethylglyoxime, the H-atom transition proceeded to the region of the radiation damage of the lattice.[139] Besides, there is only one point at the low-temperature plateau. We shall note that under tunneling in the crystal (see Section 5.3.2) there is no sharp transition to the low-temperature limit which corresponds to the results of References 72 and 73. Comparison of experimental results[42,72,73,137,139] with theory is given in Section 5.5.

One more example of the rate-constant low-temperature limit was discovered in studying the H-atom transfer in the anthracene-anthrahinon system. The reaction proceeded in the anthracene crystal added with anthrahinon. The system was pumped with powerful laser radiation (≈10 MV/cm^2). The total energy for exciting both molecules allows the hydrogen atom transit from anthracene to anthrahinon. As a result the radical pair is formed. This demands 6 eV (3.4 eV is the excitation energy of anthracene and 3 eV is that of anthrahinon).

The reserve process, the annihilation of the radical pairs, takes place at the return of the H atom to anthracene. This reaction is observed in the following manner. A small power laser excites the anthracene luminescence whose quenching is caused by the presence of the radical pair. Consequently, the intensity of luminescence increases with decreasing concentration of radical pairs. By measuring the dependence of luminescence intensity on time, it is possible to obtain the rate constant of the radical pair annihilation.

The results of measurements are presented in Figure 4.7.[141] The type of the curve describing the rate constant temperature dependence is typical for the non-Arrhenius kinetics. The temperature range is 10 to 270 K. The isotope effect in the region of the low-temperature plateau was about of the order of three.

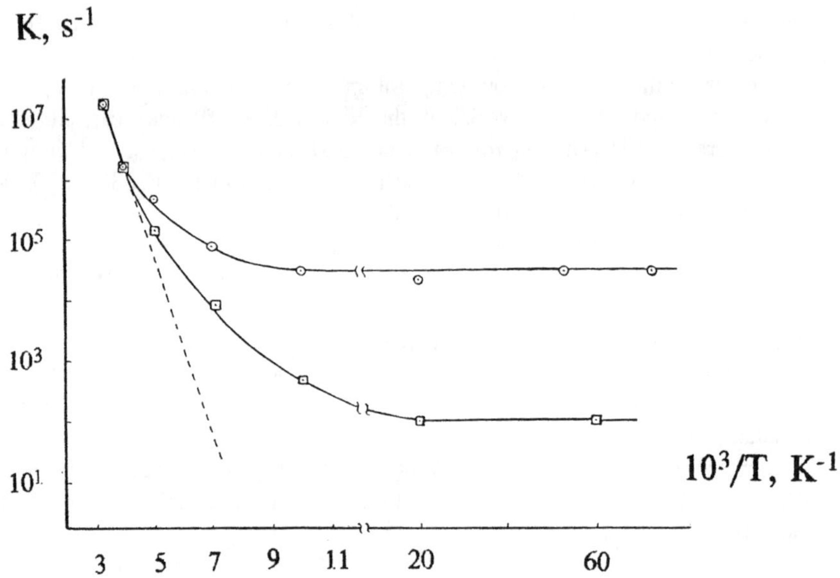

Figure 4.7 Temperature dependence of rate constant of radical pairs annihilation in anthracene-anthrahinon system.

The greater value of the isotope effect[47] is known, too. For instance, in the reaction of acetonitril molecules with methyl radical at temperature 77 K, the measured kinetic isotope effect exceeded the value $2.8 \cdot 10^4$. This is 20 times more than the value of the classical isotope effect. As is known, the classical isotope effect manifests itself as a result of the difference of energies C–H and C–D zero vibrations.

4.3.4 Low-Temperature Reactions with Transfer of Heavy Particles

Let us discuss now the possible reasons for the appearance of the rate-constant low-temperature limit with participating atoms of chlorine, bromine, and fragments of molecule.

The low-temperature rate limit of this chemical process was discovered by studying a chain hydrobromination of ethylene. The reaction was initiated under γ-radiation, with the dissociation of the HBr molecule taking place. As a result, there are two possible ways of growth chain process — either through the hydrogen atom or through the bromine atom. In spite of the fact that the H atom is lighter and at low temperatures this way of reaction seems preferable, such a development of the process is forbidden. This is associated with the fact that one of the stages is endothermic. The reaction is most likely to go a long way when the first stage is a connection of bromine atom to ethylene molecule. In this case, the reaction is exothermic with heat release in each reaction cycle of the energy equal to 0.8 eV. The kinetics of ethylene hydrobromination were studied in the range of temperatures 40 to 90 K; the low-temperature plateau started in the range 60 K.

Very interesting results were obtained in studying photochemical reactions in crystalline mixtures of ethylene with chlorine and bromine.[143] These mixtures were sprayed on the substrate, cooled up to 40 K, and then a photodissociation of halogens was conducted. Further, the halogen atom was connected to an ethylene molecule and the obtained radical started a chain reaction. The stage:

$$C_2H_4Cl + Cl_2 \rightarrow C_2H_4Cl_2 + Cl \tag{4.12}$$

was limiting. A similar process takes place in the mixture with bromine. Reactions were studied in the temperature range 20 to 70 K. The low-temperature limit in both cases at a ratio of reagents concentration of 1:1 was achieved in the range of 45 K. The rate constant temperature dependences of reactions of ethylene with bromine and chlorine turned out to be very close. The constant value in the range of the low-temperature plateau was approximately 10 sec^{-1}.

The experiment results were processed on the basis of the tunneling mechanism of the stage (4.12). A potential surface was constructed and potential barrier parameters determined for tunneling chlorine atom. They were found equal: the barrier was 1.5 kcal/mol high, 0.55 Å wide and the reduced mass of the tunneling particle 17 m_H (m_H is the mass of the hydrogen atom). The activation energy between 45 and 70 K coincided with the value E_a = 1 kcal/mol measured earlier in the gaseous phase.[144]

The experimental data obtained in this work need to be investigated further (see Section 6.5.2), since the rate constant temperature dependences for reactions with transfer of chlorine and bromine are very close. And it is clear that due to the sufficient difference of the masses of chlorine and bromine the tunneling mechanism discussed in References 10 and 143 is impossible.

After the appearance of a great many works devoted to γ-induced low-temperature polymerization of different compounds, the question arose about establishing the mechanism of this process. First, as a result of the nonselectivity of the γ-radiation effect, active centers of different natures are formed in the substance. This makes it difficult to clarify the role of each of them in the process of chain growth. Second, chains from different centers may interact. Lastly, mechanisms of polymerization for different initial substances and centers may be different. For these reasons, the study reported in Reference 145 was designed to deal with photosensitized chain polymerization of acrilonitril in the temperature range 20 to 150 K. Light acted on the admixture (TMPD), a charge transfer onto acrilonitril took place, and the obtained anion-radical was the only nucleus of the chain.

By means of infrared spectroscopy, crystalline and amorphous samples were studied. It was found that in crystalline acrilonitril only a small portion of the substance is polymerized. Up to 80 K the level of the transformation and polymerization rate does not depend on temperature; the transformations were about 10% of the amount of substance. With increasing temperature these characteristics noticeably increased. In the amorphous phase a large level of transformation was observed. The kinetics obeyed the second order law. A noticeable dependence of the polymerization kinetics of the state of the medium indicates the role of the energy stored in the substance.[111,112]

Studies of photo-induced reactions in the glassy matrix at helium temperatures include the chlorination of such saturated compounds as methylcyclohexane[146] and butyl chloride.[147] Under the effect of laser radiation, the chlorine molecule dissociated and the chlorine atom started a chain reaction. The limiting stage of these reactions is

$$R + Cl_2 \rightarrow R\text{Cl} + \text{Cl}. \qquad (4.13)$$

Here, R is the the radical which is formed from the matrix molecules (RH) in reaction with the chlorine atom:

$$R\text{H} + \text{Cl} \rightarrow R + \text{HCl}. \qquad (4.14)$$

The radicals were registered by the ESR method; chlorine molecules, spectrophotometrically in the wave length of photolysis.

In methylcyclohexane matrices the reaction was carried out in the range of temperatures 10 to 90 K. The low-temperature plateau started at 60 K. To interpret the experimental data, use was made of a model[146] suggested earlier for the tunneling of the hydrogen atom[44,45] and modulation of barrier parameters by intramolecular vibrations was taken into account. Application of this model for the transfer of a chlorine heavy atom in the elementary act (4.13) was made for the parabolic potential barrier.[146] The obtained temperature dependence of the transfer rate constant at known values of the barrier parameters does not agree with experimental data. Reference to the difference of the real shape of the potential barrier from the parabolic as the cause of nonagreement between the experiment and theory is hardly reasonable. The possible combination of the tunneling mechanism of the reaction with a transfer of a rather heavy particle (chlorine atom) causes some doubt.

Yet more differences from the process in crystal were discovered in studying the kinetics of photochlorination in the glassy matrix of bytil chlorine.[147] The range of temperatures at which the investigations were carried out was 10 to 110 K; low temperature limit was achieved in the range of 80 K. The kinetics of the process which propagates by the same scheme as that in the case of the methyl cyclohexane matrix (see stages (4.13), (4.14)) very strongly depend on the ratio of reagents concentration. The analysis of these specific features will be given in Section 6.2.3, where the influence of the thermal effect of the reaction and energy stored in the amorphous phase on the regularities of the constituent chemical properties are considered.[111,112]

4.3.5 Reactions with Participation of Hydrogen Atoms and Molecules

There are several works which discovered the reactions

$$C_2H_6 + H \rightarrow C_2H_5 + H_2 \tag{4.15}$$

at super low temperatures.[146,147] For instance, the methane matrix containing 0.5 mol% ethane at 4.2 K was exposed to γ-radiation. The ESR-method was used to observe methyl radicals, captured hydrogen atoms, and other paramagnetic particles. In the temperature range 10 to 20 K, reaction (4.15) proceeded selectively. It has been shown that this selectivity is caused by the energy difference of the dissociative bond C–H in ethane and methane molecules.[148]

In addition, specific features of (4.15) were studied in the xenon matrix. The constituent ethane molecules (0.75 mol%) in the range of temperatures 26 to 35 K react with hydrogen atoms. There atoms are produced in photolysis of molecules HI (0.15 mol%).

The rate of these reactions probably is controlled by the mobility of hydrogen atoms which has been demonstrated by investigations of the kinetic isotope effect in such systems.[149] It was found that up to a temperature of 35 K the constant rate ratio of H-atom reaction with ethane and deuterated ethane was about 2 and it was only at 50 K that this value increased up to 60. Such a regularity can be easily understood if one considers that the most time is spent transporting a hydrogen atom in the xenon matrix up to meeting an ethane molecule, and the elementary act itself does not limit the process. On the basis of the results so far reported, a conclusion is made about the tunneling mechanism of the process (4.15).[148-150]

The low-temperature reactions in alkanes were reported in Reference 151. The abstraction of the deuterium atom in the range of temperatures 5 to 29 K by medium-stabilized hydrogen and deuterium atoms was considered. Very interesting temperature dependences related most likely to the specific features of stabilization in the alkane matrix of hydrogen (deuterium) atoms.

Much attention was paid to reactions

$$H + H_2 \rightarrow H_2 + H, \tag{4.16}$$

as well as in deuterium-replaced mixtures. Process (4.16) was first observed by the methods of Reference 152. The atomic beam, containing hydrogen and deuterium, was directed after exposure of high-frequency discharge on the *He* II surface. Although the concentrations of hydrogen and deuterium atoms in discharge are approximately equal (after some time passage) the number of hydrogen atoms on the substructure is found to be considerably larger. This is caused by the fact that the reaction of the H-atom with a molecule of deuterium (due to the frequency difference of zero-molecular vibration of D_2 and HD) is endothermic and at helium temperatures does not proceed. Thus it turns out the course of the deuterium atom reaction with hydrogen molecules was demonstrated.

Such reactions of isotope exchange were further studied in Reference 153. The rate constant of the deuterium reaction with a hydrogen molecule in the range of temperatures 4.2 to 5.9 K was measured. At sample temperatures over 5.2 K the rate constant increases rather rapidly.

In more detail the tunneling reactions (4.16) were investigated in a set of works.[154-159] The atoms of hydrogen (deuterium) were formed in mixtures of molecular hydrogen with deuterium as a result of γ-radiation at temperatures 4.2 and 1.9 K. At the same temperatures the kinetics of chemical processes were investigated. As a result of the investigations a direct demonstration of the tunneling reaction was obtained

$$HD + D \rightarrow H + D_2,$$

and a significant isotope effect discovered. At temperature 4.2 K it was equal to $4 \cdot 10^4$. It is shown that the angle moment is an important parameter determining the rate constant value of these reactions.

We shall note also the studies of low-temperature monomolecular reactions.[160-162] For instance, studies have investigated the dissociation of excited radicals of hydrazine, polyvinyl alcohol, acetic anhydride, and aminoalkyl.[160] The photodissociation of the first two radicals leads to the splitting of the hydrogen atom; under excitation of acetic anhydride and aminoalkyl, respectively, methyl and ethyl radicals. A mechanism of the temperature dependence of the radical photodissociation rate constant was suggested. They were caused by interaction with neighbouring terms between which transition and medium rearrangement takes place as a result of radical dissociation. These mechanisms lead to the exponential dependence of the photodissociation rate constant.

The radiolysis of water at a temperature 4.2 K brought about a significant isotope effect in the formation of atoms of hydrogen and deuterium.[161] Hydrogen atoms (deuterium) are formed under dissociative capture of electrons by ions HD_2O^+ and D_3O^+. Thus, it is as if transformation of the captured electrons into hydrogen and deuterium atoms takes place with the value of the isotope effect larger than 10^3. Yet a larger isotope effect was observed in studying the kinetics of reaction:[162]

$$(CH_3)_2CHCH(CH_3)_2^{\ddagger} \rightarrow (CH_3)_2C = C(CH_3)_2^{\ddagger} + H_2. \quad (4.17)$$

The mixtures of compounds *h*-DMB-SF_6 and *d*-DMB-SF_6 were exposed to γ-radiation at temperature 77 K. The obtained cations decayed as result of reaction (4.17) with formation of hydrogen and deuterium molecules. The experimental value of the isotope effect turned out to be equal to $1.7 \cdot 10^4$. The quantum-chemical calculations of the isotope effect were conducted and gave the value on the order of two.

REFERENCES

1. Bell, R.P., *The Tunnel Effect in Chemistry,* Chapman & Hall, London, 1980.
2. Ovchinnikov, A.A., Ovchinnikova, M.Ya., Problems of non-linear radiationless processes in chemistry, *Adv. Quant. Chem.,* 16, 161, 1982.

3. Caldeira, A.O., Leggett, A.J., Quantum tunneling in a dissipative system, *Ann. Phys.,* 149, 374, 1983; 153, 445, 1983.
4. Jortner, J., Pullman, B., Eds., *Tunnelling,* Reidel, Dordrecht, 1986.
5. Goldanskii, V.I., Fleurov, V.N., Trakhtenberg, L.I., Contemporary concepts of the tunneling of heavy particles in chemical conversions, *Sov. Sci. Rev. B Chem.,* 9, 59, 1987.
6. Goldanskii, V.I., Trakhtenberg, L.I., Fleurov, V.N., *Tunneling Phenomena in Chemical Physics,* Gordon and Breach Science, New York, 1989.
7. Goldanskii, V.I., Benderskii, V.A., Trakhtenberg, L.I., Quantum gyrochemical reactivity of solids, *Adv. Chem. Phys.,* 75, 349, 1989.
8. Hanggi, P., Talkner, P., Borkovck, M., Rate theory: fifty years after Kramers, *Rev. Mod. Phys.,* 62, 251, 1990.
9. Trakhtenberg, L.I., Low temperatures anomalous in chemical reactions, *J. Sov. Chem. Soc.,* 35, 560, 1990.
10. Benderskii, V.A., Goldanskii, V.I., Tunneling of heavy particles in low temperature chemistry, *Intl. Rev. Phys. Chem.,* 11, 1, 1992.
11. Hanggi, P., Barrier crossing at low temperatures, in *Activated Barrier Crossing,* Fleming, G.R., Hanggi, P., Eds., World Scientific, Singapore, 1993, 268.
12. Benderskii, V.A., Makarov, D.E., Wight, C.A., *Chemical Dynamics at Low Temperatures,* John Wiley & Sons, New York, 1994.
13. Ulstrup, J., *Charge Transfer Processes in Condensed Media,* Springer-Verlag, New York, 1979.
14. Markus, R.A., Sutin, N., Electron transfer in chemistry and biology, *Biochim. Biophys. Acta.,* 811, 265, 1985.
15. Khairutdinov, R.F., Zamaraev, K.I., Zhdanov, V.P., *Electron Tunnelling in Chemistry,* Elsevier, Amsterdam, 1987.
16. Van't Hoff, J.H., *Etudes ge Dynamigues Chimigue,* F. Muller, Amsterdam, 1884, 114.
17. Arrhenius, S., Uber die reaktionsgeschwindigkeit bei der inversion von rohzucker durch sauren, *Zeitschrift f. Phys. Chemie.,* 4, 226, 1889.
18. Eyring, H., Lin, S.H., Lin, S.M., *Basic Chemical Kinetics,* John Wiley & Sons, New York, 1983.
19. Kramers, H.A., Brownian motion in a field of force and the diffusion model of chemical reactions, *Physica.,* 7, 284, 1940.
20. Eyring, H., The activated complex in chemical reactions, *J. Chem. Phys.,* 3, 107, 1935.
21. *Activated Barrier Crossing,* Fleming, G.R., Hanggi, P., Eds., World Scientific, Singapore, 1993.
22. Kim, H.J., Hynes, J.T., A theoretical model for S_N 1 ionic dissociation in solution I. Activation free energetics and transition state structure, *J. Amer. Chem. Soc.,* 114, 10508, 1992.
23. Kim, H.J., Hynes, J.T., A theoretical model for S_N 1 ionic dissociation in solution II. Nonequilibrium solvation reaction path and reaction rate, *J. Amer. Chem. Soc.,* 114, 10528, 1992.
24. Kim, H.J., Hynes, J.T., A theoretical model for S_N 1 ionic dissociation in solution III. Analysis of tret-butyl halides, *J. Amer. Chem. Soc.,* 115, 8248, 1992.
25. Truhlar, D.G., Schenter, G.K., Garrett, B.C., Inclusion of nonequilibrium continuum solvation effects in variational transition state theory, *J. Chem. Phys.,* 98, 5756, 1993.
26. Berezhkovskii, A.M., Dudko, S.A., Solvent dynamics influence of chemical reaction dynamics, *J. Chem. Phys.,* 100, 5949, 1994.
27. Berezhkovskii, A.M., Pollak, E., Zitserman, V.Yu., Activated rate process generalization of the Kramers-Grote-Hynes and Langer Theories, *J. Chem. Phys.,* 97, 2422, 1992.

28. Berezhkovskii, A.M., Frishman, A.M., Pollak, E., Variation transition state theory for multidimensional rate processes in the presence of anisotropic friction, *J. Chem. Phys.*, 101, 4778, 1994.
29. Hund, F., Bemerkungen uber das Schwingungs and Rotationsspectrum bei molekein mit mehr als zwei kernen, *Z. Phys.*, 43, 805, 1927.
30. Gamow, G.A., Zur qaantentheoric des atomkernes, *Z. Phys.*, 51, 204, 1928.
31. Bourgin, D.G., Unimolecular reactions, *Proc. Nat. Acad. Sci. U.S.A.*, 15, 357, 1929.
32. Roginsky, S.Z., Rozenkevitsch, L.V., Zum aufbay der quantentheorie der chemischen kinetik, *Z. Phys. Chem.*, B10, 47, 1930.
33. Wigner, E., Uber das uberschreiten von potentialschwellen bei chemischen reaction, *Z. Phys. Chem.*, B19, 190, 1932.
34. Bell, R.P., The application of quantum mechanics to chemical kinetics, *Proc. R. Soc. Lond. Ser.*, A139, 466, 1933.
35. Bell, R.P., Quantum mechanical effects in reactions involving hydrogen, *Proc. R. Soc. Lond. Ser.*, A148, 241, 1935.
36. Bell, R.P., Exact and approximate expressions for permeability of potential barriers to light particles, *Proc. R. Soc. Lond. Ser.*, A158, 128, 1937.
37. Bell, R.P., *The Proton in Chemistry*, 2nd ed., Cornell University Press, New York, 1973.
38. Goldanskii, V.I., The role of tunnel effect in the kinetics of chemical reactions at low temperatures, *Dokl. Phys. Chem.*, 124, 1261, 1959.
39. Goldanskii, V.I., Tunnel transitions between systems described by Morse potential curves, *Dokl. Phys. Chem.*, 127, 1037, 1959.
40. Yakimchenko, O.E., Lebedev, Ya.S., Radical pairs in irradiated dimethylglyoxine isotope and tunnel effects in pairs transformations, *Intl. J. Radiat. Chem.*, 3, 17, 1971.
41. Sprague, E.D., Williams, F., Evidence for hydrogen atom abstraction by methyl radicals in the solid state at 77 K, *J. Amer. Chem. Soc.*, 93, 787, 1971.
42. Le Roy, R.J., Murai, H., Williams, F., Tunneling model for hydrogen abstraction reactions in low-temperature solids. Applications to reactions in alcohol glasses and acetonitril crystals, *J. Amer. Chem. Soc.*, 102, 2325, 1980.
43. Jortner, J., Ulstrup, J., Tunneling in low-temperature atom-transfer processes, *Chem. Phys. Lett.*, 63, 236, 1979.
44. Klochikin, V.L., Pshezhetskii, S.Ya., Trakhtenberg, L.I., On the theory of free radicals recombination in solid phase, *Dokl. Phys. Chem.*, 239, 324, 1978.
45. Klochikhin, V.L., Pshezhetskii, S.Ya., Trakhtenberg, L.I., The tunneling mechanism of free valences mobility in polymers, *Vysokol. Soedin.*, A21, 2792, 1979.
46. Klochikin, V.L., Pshezhetskii, S.Ya., Trakhtenberg, L.I., On the temperature dependence of the rate constants for the reactions with hydrogen atom tunneling, *Khim. Visokikh. Energii.*, 14, 329, 1980.
47. Sprague, E.D., Deuterium isotope effect on hydrogen atom abstraction by methyl radicals in acenitrile at 77 K. Evidence for tunneling, *J. Phys. Chem.*, 81, 516, 1977.
48. Marcus, R.A., On the theory of oxidation-reduction reactions involving electron transfer, I., *J. Chem. Phys.*, 24, 966, 1956.
49. Marcus, R.A., On the theory of oxidation-reduction reactions involving electron transfer, II. Applications of data on the rate of isotopic exchange reactions, *J. Chem. Phys.*, 26, 867, 1959.
50. Dogonadze, R.R., Kuznetsov, A.M., Levich, V.G., Theory of hydrogen atom discharge on metals: case of overvoltages, *Electrochim. Acta*, 13, 1025, 1968.
51. Kuznetsov, A.M., Recent advances in the theory of charge transfer, *Mod. Aspects Electrochem.*, 20, 95, 1989.
52. Klochikhin, V.L., Trakhtenberg, L.I., Effect of crystal lattice oscillations on the tunneling rate of atoms in solid phase, *Dokl. Phys. Chem.*, 259, 399, 1981.

53. Benderskii, V.A., Goldanskii, V.I., Ovchinnikov, A.A., Effect of molecular motion on low-temperature and other anomalously fast chemical reactions in the solid phase, *Chem. Phys. Lett.*, 73, 492, 1980.
54. Johnston, H.S., Rapp. D., Large tunneling corrections in chemical reaction rates, *J. Amer. Chem. Soc.*, 83, 1, 1961.
55. Truhlar, D.G., Kupperman, A., Exact tunneling calculations, *J. Amer. Chem. Soc.*, 93, 1840, 1971.
56. Miller, W.H., Path integral representation of the reaction rate constant in quantum mechanical, transition state theory, *J. Chem. Phys.*, 63, 1166, 1975.
57. Ovchinnikova, M.Ya., The tunneling dynamics of the low temperature H exchanage reactions, *Chem. Phys.*, 36, 85, 1979.
58. Babamov, V.K., Marcus, R.A., Dynamics of hydrogen atom and proton transfer reactions. Symmetric case, *J. Chem. Phys.*, 74, 1790, 1981.
59. Babamov, V.K., Lopez, V., Marcus, R.A., Dynamics of hydrogen atom and proton transfer reactions. Nearly degenerate asymmetric case, *J. Chem. Phys.*, 78, 5621, 1983.
60. Klochikhin, V.L., Pshezhetskii, S.Ya., Trakhtenberg, L.I., Tunneling transition of atom from the molecule to the radical with energy transfer in vibrations of the "radical-molecule" complex, *Russ. J. Phys. Chem.*, 54, 1324, 1980.
61. Trakhtenberg, L.I., Klochikhin, V.L., Pshezhetskii, S.Ya., Tunneling of hydrogen atom in low temperature processes, *Chem. Phys.*, 59, 191, 1981.
62. Trakhtenberg, L.I., Klochikhin, V.L., Pshezhetskii, S.Ya., The theory of the tunnel transition of atoms in solids, *Chem. Phys.*, 69, 121, 1982.
63. Flomenblit, V.Sh., Trakhtenberg, L.I., The temperature dependence of the rate constant of a hydrogen atom transfer in a solid-state reaction within a fluorene-acridine system, *Sov. J. Chem. Phys.*, 11, 1324, 1992.
64. Ivanov, G.K., Kozhushner, M.A., Nonadiabatic effects in tunneling reactions of heavy particles, *Sov. J. Chem. Phys.*, 2, 1299, 1983.
65. McKinon, W.R., Hurd, C.M., Tunneling and the temperature dependence of hydrogen transfer reactions, *J. Chem. Phys.*, 87, 1283, 1983.
66. Brunton, G., Griller, D., Barclay, L.R.C., Ingold, K.U., Kinetic applications of electron paramagnetic resonance spectroscopy. 26. Quantum-mechanical tunneling in the isomerization of sterically hindered aril radicals, *J. Amer. Chem. Soc.*, 98, 6803, 1976.
67. Brunton, G., Gray, J.A., Griller, D., Barclay, L.R.C., Ingold, K.U., Kinetic applications of electron paramagnetic resonance spectroscopy. 32. Further studies of quantum-mechanical tunneling in the isomerization of sterically hindered aril radicals, *J. Amer. Chem. Soc.*, 100, 4197, 1978.
68. Grellmann, K.H., Schmitt, U., Weller, H., Tunnel effect on the kinetics of a sigmatropic proton shift, *Chem. Phys. Lett.*, 88, 40, 1982.
69. Siebrand, W., Wildman, T.A., Zgierski, M.Z., Temperature dependence of hydrogen tunneling rate constants, *Chem. Phys. Lett.*, 98, 108, 1983.
70. Chantranupong, L., Wildman, T.A., Golden-rule treatment of hydrogen abstraction by photoexcited acridine guests in fluorene single crystals, *J. Chem. Phys.*, 94, 1030, 1991.
71. Lavtchieva, L., Smedarchina, Z., Generalized golden rule treatment of a photochemical solid-state reaction involving hydrogen tunneling, *Chem. Phys. Lett.*, 184, 545, 1991.
72. Prass, B., Colpa, J.P., Stehlik, D., Identification of the lowest energy nuclear fluctuation modes promoting the photochemical H-transfer tunneling reaction in doped fluorene single crystals, *J. Chem. Phys.*, 88, 191, 1988.
73. Prass, B., Colpa, J.P., Stehlik, D., Intermolecular H-tunneling in a solid-state promoted by distinct low-energy nuclear fluctuation modes, *Chem. Phys.*, 136, 187, 1989.

74. Colpa, J.P., Prass, B., Stehlik, D., Evidence for tunneling as a mechanism for photochemical hydrogen-transfer reaction in molecular crystals, *Chem. Phys. Lett.,* 107, 469, 1984.
75. Prass, B., Fujara, F., Stehlik, D., Kinetics of photochemical H-abstraction by $^3\pi\pi^*$ acridin fluorene single crystals as studied by triplet-triplet absorbtion, *Chem. Phys.,* 81, 175, 1983.
76. Gribb, P.H., Resonance enhanced tunneling in an asymmetric double well potential: the $2B^1\Sigma^+$ state of HCl, *Chem. Phys.,* 88, 47, 1984.
77. Belorizky, E., Fries, P.H., Intermolecular tunneling between diffusing spherical potential wells; a first approach to direct proton transfer in solution, *J. Phys. Fr.,* 49, 727, 1988.
78. Borgis, D.C., Lee, S., Hynes, J.T., A dynamical theory of nonadiabatic proton and hydrogen atom transfer reaction rate in solution, *Chem. Phys. Lett.,* 162, 19, 1989.
79. Suarez, A., Silbey, R., Hydrogen tunneling in condensed media, *J. Chem. Phys.,* 94, 4809, 1991.
80. Borgis, D.C., Hynes, J.T., Molecular-dynamics simulation for a model nonadiabatic proton transfer reaction in solution, *J. Chem. Phys.,* 94, 3619, 1991.
81. Smedarchina, Z., Generalized golden rule approach to the H-transfer problem in poliatomic systems. I. Multimode nuclear effects in solids, *Chem. Phys.,* 150, 47, 1991.
82. Chandra, A.K., Sreedhara Rao, V., Temperature dependence of hydrogen transfer reactions via tunneling at low temperatures, *Intl. J. Quant. Chem.,* 47, 437, 1994.
83. Caldeira, A.O., Leggett, A.J., Influence of dissipation on quantum tunneling in macroscopic systems, *Phys. Rev. Lett.,* 46, 211, 1981.
84. Caldeira, A.O., Leggett, A.J., Path integral approach to quantum Brownian motion, *Physica A (Utrecht),* 121, 587, 1983; 130, 374, 1983 (Errtum). ·
85. Leggett, A.J., Chakravarty, S., Dorsey, A.T., Fisher, M.P.A., Garg, A., Dynamics of dissipative two-state system, *Rev. Mod. Phys.,* 59, 1, 1987.
86. Levine, A.M., Hontscha, W., Pollak, E., Sudden theory for tunneling in dissipative systems, *Phys. Rev.,* B40, 2138, 1989.
87. Rips, I., Pollak, E., Kramers model: solution of the turnover problem, *Phys. Rev.,* A41, 5366, 1990.
88. Tanimura, Y., Wolynes, P.G., The interplay of tunneling resonance and dissipation in quantum barrier crossing: a numerical study, *J. Chem. Phys.,* 96, 8485, 1992.
89. Topaler, M., Makri, N., Quantum rates for a double well coupled to a dissipative bath: accurate path integral results and comparison with approximate theories, *J. Chem. Phys.,* 101, 7500, 1994.
90. Bikbayev, N.Kh., Ivanov, A.I., Lomakin, G.S., Ponomarev, O.A., Theory of radiationless process in the "non-Condon" approximation, *Sov. Phys. JETF,* 74, 2154, 1978.
91. Trakhtenberg, L.I., Effect of the medium on the rate of intramolecular radiationless transitions, *Dokl. Phys. Chem.,* 269, 1131, 1983.
92. Shida, N., Barbara, P.F., Almloff, J., A reaction surface Hamiltonian treatment of the double proton transfer of formic acid dimer, *J. Chem. Phys.,* 94, 3633, 1991.
93. Shida, N., Almloff, J., Barbara, P.F., Tunneling paths in intramolecular proton transfer, *J. Phys. Chem.,,* 95, 10457, 1991.
94. Li, D., Voth, G.A., Feyman path integral approach for studying intramolecular effects in proton-transfer reactions, *J. Phys. Chem.,* 95, 10425, 1991.
95. Sokolov, N.D., Vener, M.V., Proton tunneling assisted by the intermolecular vibration excitation in solid state, *Chem. Phys.,* 168, 29, 1992.
96. Luth, K., Scheiner, S., Excited-state energetics and proton transfer barriers in malonaldegide, *J. Phys. Chem.,* 98, 3582, 1994.

97. Smedarchina, Z., Enchev, V., Lavchieva, L., Theoretical study of intramolecular proton transfer in solution: application to the photoenolization of 5,8-dimethyl-1-tetralone, *J. Phys. Chem.,* 98, 4218, 1994.
98. Klochikhin, V.L., Trakhtenberg, L.I., Effect of intermolecular vibration on the charge transfer rate in solid phase, *Sov. J. Electrochem.,* 19, 1212, 1983.
99. Trakhtenberg, L.I., Effect of medium on chemical reactions rate in solid phase, *Sov. J. Chem. Phys.,* 2, 1683, 1983.
100. Takada, S., Nakamura, H., Wentrel-Kramers-Brillouin theory of multidimensional tunneling: general theory for energy splitting, *J. Chem. Phys.,* 100, 98, 1994.
101. McCoy, A., Gerber, R.B., Rather, M.A., A quantitative approximation for the quantum dynamics of hydrogen transfer: transition state dynamics and decay in $ClHCl^-$, *J. Chem. Phys.,* 101, 1975, 1994.
102. Metz, R.B., Kitsopoulos, T., Weaver, A., Neumark, D.M., Study of the transition state region in the Cl + HCl reaction by photoelectron spectroscopy of $ClHCl^-$, *J. Chem. Phys.,* 88, 1463, 1988.
103. Flynn, C.P., Stoneham, A.M., Quantum theory of diffusion with application to light interstitials in metals, *Phys. Rev.,* B1, 3966, 1970.
104. Kagan, Yu., Klinger, M.I., Role of fluctuational "preparation" of barriers in quantum diffusion of atomic particles, *Sov. Phys. JETF,* 70, 255, 1976.
105. Johnston, H.S., Large tunneling correction in chemical reaction rates, *Adv. Chem. Phys.,* 3, 131, 1960.
106. De Vault, D., Chance, B., Photosynethesis using a pulsed laser, I. Temperature of cytochrome oxidation rate in chromatium. Evidence for tunneling, *Biophys. J.,* 6, 825, 1966.
107. Zamaraev, Ê.I., Khairutdinov, R.F., Mikhailov, À.I., Goldanskii, V.I., Tunneling effect in reactions of electron transfer in condensed phase, *Dokl. Phys. Chem.,* 199, 640, 1971.
108. Bruk, V.A., Gromov, V.F., Chernyak, I.V., Khomikovskii, P.M., Abkin, A.D., À Ràdiation polymerization of tetrafluorethylene and acrylonitril at 4,2Ê, *Polym. Sci.,* 8, 961, 1966.
109. Barkalov, I.M., Gareeva, D.A., Goldanskii, V.I., Enikolopyan, I.E., Berlin, A.A., Ràdiation polymerization of monomers irradiated at 4,2K, *Polym. Sci.,* 8, 1140, 1966.
110. Goldanskii, V.I., Chemical reaction at very low temperatures, *Ann. Rev. Phys. Chem.,* 27, 85, 1976.
111. Timashev, S.F., Intermittency in chemical reactions kinetics in solids, *Russ. J. Phys. Chem.,* 66, 846, 1992.
112. Timashev, S.F., Trakhtenberg, L.I., Classical above barrier mechanism of the low temperature limit of chemical reaction rate, *Russ. J. Phys. Chem.,* 67, 401, 1993.
113. Goldanskii, V.I., Frank-Kamenskii, M.D., Barkalov, I.M., Quantum low temperature limit of a chemical reaction rate, *Science,* 182, 1344, 1973.
114. Basilevskii, M.V., Gerasimov, G.N., Petrochenko, S.I., Calculation on the mechanism of ion polymerization in the formaldehyde crystal, *Chem. Phys.,* 72, 349, 1982.
115. Punkkinen, V., Temperature dependence of liberation tunneling and proton relaxation, *Phys. Rev.,* B21, 54, 1980.
116. Kiryukhin, D.P., Barkalov, I.M., Goldanskii, V.I., Post-polymerization of formaldehyde at 5 K, *Polym. Sci.,* 16B, 565, 1974.
117. Goldanskii, V.I., Facts and hypotheses of molecular chemical tunneling, *Nature,* 279, 109, 1979.
118. Alberding, N., Austin, R.H., Chan, S.S., Eisenstein, L., Frauenfelder, H., Gunsalus, I.C., Nordlund, T.M., Dynamics of carbon monoxide billing to protoheme, *J. Phys. Chem.,* 65, 4701, 1976.

119. Alberding, N., Austin, R.H., Beenson, K.W., Chan, S.S., Eisenstein, L., Frauenfelder, H., Nordlund, T.M., Tunneling in ligand binding to heme proteins, *Science,* 192, 1002, 1976.
120. Alben, J.O., Beece, D., Browne, C.F., Eisenstein, L., Frauenfelder, H., Good, D., Marden, M.C., Moh, P.P., Reinisch, L., Reynolds, A.H., Yue, K.I., Isotope effect in molecular tunneling, *Phys. Rev. Lett.,* 44, 1157, 1980.
121. Plonka, A., *Lecture Notes in Chemistry,* Springer-Verlag, Berlin, 1989.
122. Frauenfelder, H., Sligar, S.G., Wolynes, P.G., The energy landscapes and motion of proteins, *Science,* 254, 1598, 1991.
123. Bohz, H., Wang, J., Wolynes, P., Growth of domains in distance geometry through protein folding, in *Protein Structure by Distance Analysis,* Hohz, H., Brunak, S., Eds., IOS Press, Amsterdam, 1994, 98.
124. Buchwalter, S.L., Closs, G.L., Electron spin resonance and CIDNP studies on 1,3-cyclopentadiyls. A localized 1,3-carbon biradical system with a triplet ground state. Tunneling in carbon-carbon bond formation, *J. Amer. Chem. Soc.,* 101, 4688, 1979.
125. Lucas, D., Pimentel, G.C., Reaction between nitric oxide and ozone in solid nitrogen, *J. Phys. Chem.,* 83, 2311, 1979.
126. Hallam, H.E., *Vibrational Spectroscopy of Trapped Species,* John Wiley & Sons, New York, 1973.
127. Bartelt, G., Eychmuller, A., Grellman, K.H., Experimental studies of hydrogen tunneling. Utilization of large isotope effect in mechanistic study of a hydrogen shift reaction, *Chem. Phys. Lett.,* 118, 568, 1985.
128. Grellman, K.H., Weller, H., Tauer, E., Tunnel effects of the kinetics of hydrogen shifts. The enol-keton transformation of 2′-methylacetophenon, *Chem. Phys. Lett.,* 95, 195, 1983.
129. Tokumura, K., Watanabe, Y., Ito, M., Deuterium isotope effect of excited-state and ground-state double-proton-transfer processes of the 7-asaindole H-bonded dimer in 3-methylpentane, *J. Chem. Phys.,* 90, 2362, 1986.
130. Grellman, K.H., Mordzinski, A., Heinrich, A., Intramolecular proton transfer and tunnel effects in the metastable triplet states of 2-(2′-hydroxyphenyl)benzoxazole studied by micro- and nanosecond laser flash photolysis, *Chem. Phys.,* 136, 201, 1989.
131. Deycard, S., Lusztyk, J., Ingold, K.U., Zerbetto, F., Zgierski, M.Z., Siebrand, W., Inversion of the oxiranyl radical occurs by quantum-mechanical tunneling, *J. Amer. Chem. Soc.,* 110, 6721, 1988.
132. Deycard, S., Lusztyk, J., Ingold, K.U., Zerbetto, F., Zgierski, M.Z., Siebrand, W., Inversion of the dioxiranyl radical in experimental and theoretical studies, *J. Amer. Chem. Soc.,* 112, 4284, 1990.
133. Al-Soufi, W., Grellmann, K.H., Nickel, B., Keto-enol tautomerization of 2-(2′-hydroxyphenyl)benzoxazole in the triplet state: hydrogen tunneling and isotope effects. 1. Transiet absorption kinetics, *J. Phys. Chem.,* 95, 10503, 1991.
134. Eisenberger, H., Nickel, B., Ruth, A.A., Al-Soufi, W., Grellmann, K.H., Novo, M., Keto-enol tautomerization of 2-(2′-hydroxyphenyl)benzoxazole and 2-(2′-hydroxy-4′-methylphenyl)benzoxazole in the triplet state: hydrogen tunneling and isotope effects. 2. Duel phosphorescence kinetics, *J. Phys. Chem.,* 95, 10509, 1991.
135. Kensy, U., Mosquera Gonzales, M., Grellman, K.H., Tunnel effects on inter- and intramolecular hydrogen transfer reactions of transient dihydro- and hexhydrocarbazoles, *Chem. Phys.,* 170, 381, 1993.
136. Kensy, U., Gonzales, M., Grellmann, K.H., Mosquera Gonzales, M., Competition between tunnel and viscosity effects on bimolecular hydrogen-transfer reaction, *Pure Appl. Chem.,* 65, 165, 1993.

137. Hudson, R.L., Shiotany, M., Williams, F., Hydrogen atom abstraction by methyl radical in methanol glasses at 15-100 K. Evidence for a limiting rate constant below 40 K by quantum-mechanical tunneling, *Chem. Phys. Lett.*, 48, 193, 1977.
138. Le Roy, R.J., Sprague, E.D., Williams, F., Quantum mechanical tunneling in hydrogen atom abstraction from solid acetonitrile at 77-87 K, *J. Phys. Chem.*, 76, 546, 1972.
139. Toriyama, K., Nunome, K., Iwasaki, M., Evidence for tunneling hydrogen atom transfer reactions, at 402 K in organic crystals, *J. Amer. Chem. Soc.*, 99, 5823, 1977.
140. Becker, R.S., *Theory and Interpretation of Fluorescence and Phosphorescence*, Wiley-Interscience, New York, 1969.
141. Private information from Benderskii, V.A., Regarding the results obtained by Benderskii, V.A. and Lavrushko, A.G. (see also References 10 and 12).
142. Barkalov, I.M., Goldanskii, V.I., Kiryukhin, D.P., Zanin, A.M., Kinetics and mechanism of the low temperature chain hydrobromination of ethylene, *Chem. Phys. Lett.*, 73, 273, 1980.
143. Benderskii, V.A., Goldanskii, V.I., Makarov, D.E., Misochko, E.Ya., Dynamics of multidimensional nuclear tunneling in the chain-ethylene chlorinate at liquid-helium temperature, *Chem. Phys. Lett.*, 179, 334, 1991.
144. Chiltz, G., Goldfinger, P., Huybrechts, G., Martens, G., Verbeke, G., Atomic chlorination of simple hydrocarbon derivatives in the gas phase, *Chem. Rev.*, 63, 355, 1963.
145. Gerasimov, G.N., Dolotov, S.M., Abkin, A.D., Solid-state photosensitized polymerization of acrylonitrile at cryogenic temperature, *Radiat. J. Chem. Phys.*, 15, 405, 1980.
146. Filippov, P.G., Misochko, E.Ya., Kulikov, A.V., Benderskii, V.A., Ovchinikov, A.A., Photochlorination of glasses matrix of the saturated compounds at helium temperatures, *Dokl. Phys. Chem.*, 256, 1173, 1981.
147. Benderskii, V.A., Misochko, E.Ya., Filippov, P.G., Kinetics peculiarities of the chain chemical reactions in the glass region of the reagents mixture, *Sov. J. Chem. Phys.*, 5, 955, 1986.
148. Iwasaki, M., Toriyama, K., Muto, H., Nunome, K., Hydrogen abstraction by thermal hydrogen atoms at 10-30 K: electron spin resonance evidence, *Chem. Phys. Lett.*, 56, 494, 1978.
149. Toriyama, K., Nunome, K., Iwasaki, M., Reactions of thermal hydrogen atoms at cryogenic temperature below 77 K as studied by ESP. Isotope effect in hydrogen abstraction from ethane in xenon matrices, *J. Phys. Chem.*, 84, 2374, 1980.
150. Iwasaki, M., Toriyama, K., Nunome, K., Fukaya, M.K., Mujto, H., Effect of irradiation temperatures on hydrogen atom reactions in neopentane and its mixtures irradiated at 4.2 and 77 K as studied by electron spin resonance, *J. Phys. Chem.*, 81, 1410, 1977.
151. Wang, H.Y., Willard, J.E., Effect of molecular structure and temperature on the tunneling abstraction of D from C–D bonds by trapped hand D atoms in alkane glasses at 5-29 K, *J. Phys. Chem.*, 83, 2585, 1979.
152. Gordon, E.B., Pelmenev, A.A., Pugachev, O.F., Khmelenko, V.V., Hydrogen and deiterium atoms stabilized by condensation of atom beam in superfluidity helium, *JETP Lett.*, 37, 237, 1983.
153. Ivliev, A.V., Iskovskikh, A.S., Katunin, A.Ya. et al., Investigation of kinetics behaviour of hydrogen and deiterium atoms in solid solution of H_2 and D_2 at helium temperatures, *JETP Lett.*, 38, 317, 1983.
154. Tsuruta, H., Miyazaki, T., Fueki, K., Azuma, N., Remarkable isotope effect on production and decay of D and H atoms in γ-radiolysis of D_2-H_2 mixtures at 4 K. A quantum-mechanical tunneling effect, *J. Phys. Chem.*, 87, 5422, 1983.

155. Miyazaki, T., Lee, K.-P., Fueki, K., Takeuchi, A., Temperature effect on the decay of H(D) atoms in the radiolysis of solid H_2, D_2 and HD at 4.2 K and 1.9 K. Evidence for tunneling migration, *J. Phys. Chem.,* 88, 4959, 1984.
156. Miyazaki, T., Lee, K.-P., Direct evidence for tunneling reaction $HD+D \rightarrow H+D_2$ in the radiolysis of D_2-HD mixtures at 4.2 K, *J. Phys. Chem.,* 90, 401, 1986.
157. Lee, K.-P., Miyazaki, T., Fueki, K., Gotoh, K., Rate constant for tunneling reaction $D_2+D \rightarrow D+D_2$ in radiolysis of D_2-HD mixtures at 4.2 K and 1.9 K, *J. Phys. Chem.,* 91, 180, 1987.
158. Miyazaki, T., Iwata, N., Lee, K.-P., Fueki, K., Decay of H(D) atoms in solid hydrogen at 4.2 K. Rate constant for tunneling reaction $H_2(D_2,HD)+H(D)$, *J. Phys. Chem.,* 93, 3352, 1989.
159. Miyazaki, T., Kitumura, S., Morikita, H., Fueki, K., Rate constant for tunneling reaction $H_2+D \rightarrow H+HD$ in the solid D_2-H_2 mixtures at 4 K, *J. Phys. Chem.,* 96, 10331, 1992.
160. Trakhtenberg, L.I., Slavinskaya, N.A., Pshezhetskii, S.Ya., The effect of dynamic properties of the medium on the kinetics solid-state radical photodissociation processes, *Chem. Phys.,* 134, 127, 1989.
161. Muto, H., Matsuura, K., Nunome, K., Large isotope effect due to quantum tunneling in the conversion reaction of electron to H and D atoms in irradiated H_2O/D_2O ice, *J. Phys. Chem.,* 96, 5211, 1992.
162. Miyazaki, T., Kitumura, S., Kozotono, Y., Matsunaga, H., A large isotope effect on formation of tetramethylethylene cations in γ-irradiated 2,3-dimethylbutane at 77 K. Tunneling of H_2 molecule, *J. Phys. Chem.,* 98, 10767, 1994.

5 MODIFIED THEORY OF RADIATIONLESS TRANSITIONS

This chapter gives a brief but consistent presentation of the mathematical means of describing solid-phase radiation-chemical reactions.

First of all, selection of a model is carried out and on its basis the expression of a chemical process rate constant is obtained. In doing so, a double adiabatic approximation is used which enables not only the separation of the motion of atoms and electrons but also the separation of inter- and intramolecular subsystems in the atomic motion.

The formula for the rate constant is found, allowing the patterns of the atomic particles transfer to be described in a wide range of temperatures — from helium to room temperatures. Although the reaction is investigated in a general form when the tunneling particle is not defined concretely, the approximations used are primarily correct for the H-atom transfer.

The rate constant expression is analyzed for various models describing the intermolecular motion of the medium. The Debye and Einstein models of the solid-phase lattice vibration are considered. Furthermore, note is taken of the fact that in the irradiated matrix one of the reagents, as a rule, is a radical. Usually this radical differs noticeably from the molecules of the medium in its mass and parameters of interaction with the environment, i.e., it is a defect. Since one of the reagents is a defect, then to obtain the expression for the rate constant it is necessary to take into account the changes in the spectrum and the dynamics of vibrations even in the case when the total number of radicals (or other radiation damages) is not too large. The role of the orientational intermolecular vibrations in radical solid-phase reactions is discussed, too.

Some results of the investigation of atomic particle transfer in the condensed phase are cited. Special attention is paid to the influence of the medium rearrangement on the transition rate constant. The chapter concludes with a comparison of theory and experimental data in the literature.

5.1 GENERAL

As is known, both the tunneling and above-barrier transfer of a particle are influenced by parameters of the barrier overcome. To make it definite let us begin with consideration of tunneling that prevails at relatively low temperatures. The theory behind the mathematics used is that under an increase in temperature the rate constant acquires the usual activation form of Equation (4.9).

According to modern concepts[1,2] the most important mechanism of the medium effect on the atomic transfer is contained in a viscous friction and in the changes to parameters of the potential barrier overcome under the intermolecular vibrations of reagents.

The account of viscosity in tunneling processes allows the energy dissipation to be accounted for; which in a most essential way manifests itself at super low temperatures in the region of absolute zero (see also Section 4.3.2, in which the role of friction is discussed).[3,4] Experimental investigations of low-temperature chemical reactions were carried out, as a rule, at some higher temperatures[1,2,5] and thus one can come to a conclusion that it is not the friction that determines their kinetics. At the same time, the intermolecular vibrations, as will be shown below, change the efficient potential barrier and, consequently, the probability of transition in the whole range of temperatures. At low temperatures, zero vibrations of reagents play a part in a temperature increase, the thermal ones.

Thus, in obtaining the rate constant of a chemical process we shall consider only a change in the reaction potential barrier under the vibrations of reagents and neglect the viscosity of the medium. Furthermore, only reactions with a light atom transfer are considered and the transfer of heavier particles will be studied separately. In conclusion, we note that a detailed treatment of tunneling on the basis of radiationless transition theory is given in Reference 1. Here, we shall dwell briefly on general aspects in connection with interpretation of the experimental data.

5.1.1 Model of Solid-Phase Chemical Reaction

Let us consider particle behavior in the two-well potential (Figure 5.1). Such a potential can correspond to the radical reaction:

$$R - A + R' \rightarrow R + A - R. \tag{5.1}$$

Here, R, R' are the free radicals; A is the tunneling atom. At the initial moment of a time, the particle A is found in the left-hand potential well and it then performs a transition to the right-hand well.

As is known, the probability density to find out the particle at some instant of time at a certain point of the space is determined from the Schrodinger equation:

$$\frac{\partial \psi(r,t)}{\partial t} = -\frac{\hbar^2}{2m}\Delta\psi(r,t) + U(r)\psi(r,t), \tag{5.2a}$$

$$\int_{b_1}^{b_1'} dr\left|\psi(r,0)\right|^2 = 1. \tag{5.2b}$$

Here, m is the mass of the tunneling particle, r is its coordinate; $U(r)$ is the potential energy of the tunneling particle. The ratio (5.2b) corresponds to the initial condition

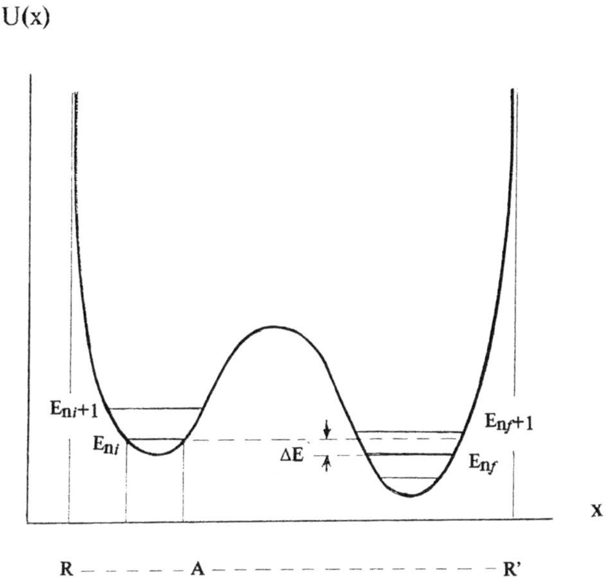

Figure 5.1 Double-well potential, corresponding to radical reaction $R–A + R' \rightarrow R + A–R'$. Indexes n_i, n_f are the number of states in the left and right wells; ΔE is the energy defect of the particle in initial and final states.

according to which at $t = 0$ the particle is localized in the left-hand potential well; $|\psi(r,t)|^2$ is the probability density; b_1, b_1' is the boundary of the particle localization in the left-hand potential well.

If one knows the potential $U(r)$ of the system studied, then in principle and having solved Equation (5.2a) the desired probability density can be found. For reaction (5.1) where the atom performs a transition from the potential curve describing the chemical coupling R-A (Figure 5.2a) over the curve corresponding to the coupling R-R' (Figure 5.2b), different forms of the potential barrier are possible (this problem will be discussed below) including that of Figure 5.1. For clarity, let us begin the discussion of a particle transfer process in this potential.

The exact solution of Schrodinger equation (5.2a) with initial condition (5.2b) unfortunately does not exist. Moreover, to determine an approximate solution of this equation also is not an easy task. In the given case, the standard technique of quantum mechanics that was developed for studying transitions under the effect of disturbances is not admissible.[6]

Indeed, the reason for a transition is the presence of a second potential well rather than the external factor. In the final state when the particle is localized in the second well there is a probability different from zero of finding it in the first potential well. Hence, the disturbance is not only small (the depths of the potential wells are commensurate) but also different for the initial and the final states. Apart from this fact, the wave functions corresponding to Schrodinger equations with potential energies of Figure 5.2a and b are not orthogonal, which involves additional problems.

In spite of the objective difficulties, they are quite solvable if the two-level approximation holds true. The applicability of the two-level approximation is ensured by the smallness of the matrix transition elements in comparison with the

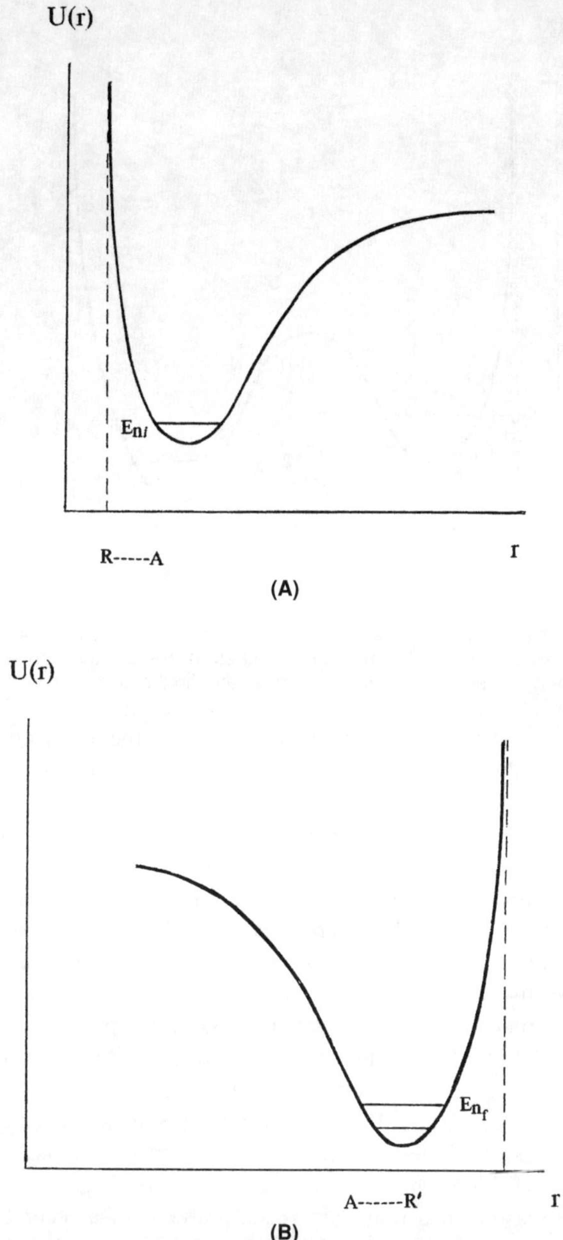

Figure 5.2 (A) Potential energy, corresponding to chemical bond *R–A*. (B) Potential energy, corresponding to chemical bond *A–R′*.

spacing between the energetic levels in each potential well. In this case, the wave function corresponding to the Hamiltonian with the potential energy shown in Figure 5.1 can be presented as a linear combination of the wave functions corresponding to the potentials shown in Figure 5.2a and b, while the differential equation of partial derivatives (5.2a) transforms into a system of two linear algebraic equations. The solution of these equations is the oscillating function of time, and, consequently, the particle coherently vibrates between two potential wells. The frequency of these vibrations depends on the energy defect (ΔE) and the matrix transition element, if one of these parameters is small, it is proportional to the second.

5.1.2 Relaxation Processes

The process of a particle transfer becomes irreversible only if the relaxation processes are taken into account. The relaxation can be due to the fact that the state into which a particle transits is not ground and its lifetime is finite with respect to the transitions to the lower vibrational levels. The rate of such transitions is rather high and can noticeably exceed the frequency of tunneling oscillations. Another type of relaxation also should be accounted for. This relaxation leads to the loss of information about the phase between the initial and the final states of the system and is caused by interaction of the two-level system with the medium. Such an interaction can also reduce the lifetime of the final state with respect to transitions to the lower levels. It should be noted that the interaction with the medium cannot be taken on the basis of the Schrodinger equation (5.2a); it is only possible using a density matrix formulation.

Thus, it can be concluded that the method of tunneling description from the bound to the unbound state depends on the relation between the time of quantum oscillations and the time of relaxation. If the relaxation time exceeds the time of oscillations, then the transfer of the particle is coherent and the probability of finding the particle in this or that potential well represents the oscillating function of time. At relaxation times less than the period of quantum oscillations, we deal with an incoherent process that is described by conventional classical kinetic equations. In a case when these times are comparable, the tunneling kinetics is described by means of density matrix mathematics.

It is interesting that all the above-mentioned parameter relationships can be realized in one system, since the oscillation period of a particle and the constant relaxations are different functions of temperature. For instance, on decreasing a temperature we suppress the relaxation and the incoherent transfer becomes coherent.

5.1.3 Potential Energy of the Tunneling Particle

There are several cases when the coherent transfer cannot be realized even at super low temperatures. First, one should take into account that in the solid phase there is always a phonon subsystem with a continuous spectrum of oscillations. If the energy defect (resonance disturbance) exceeds the magnitude of the transition matrix element, then the dissipation of energy released into the phonon subsystem provides tunneling incoherence. Second, in chemical kinetics, reactions are often encountered where the final state of the intramolecular vibrational subsystem belongs to the continuous spectrum. These are reactions with a large thermal effect when the reaction products are excited vibrationally. (In compound molecules and especially in those interacting with the environment, the continuous spectrum in the oscillating subsystem begins with third excited level.) In addition, such processes include the dissociation of electron-excited molecules and radicals.

In all cases when the particle transition is incoherent, the tunneling rate constant can be written by means of Fermi's "Golden rule" law:

$$k = \frac{2\pi}{\hbar} Av_i \sum_f |\langle f|V|i\rangle|^2 \delta\left(E_f - E_i\right). \tag{5.3}$$

Here the summation (Σ_f) is performed over all final states and averaging (Av_i) over the initial states of the system; V is the disturbance operator causing the tunneling; $|i\rangle$, $|f\rangle$ are the wave functions of the initial and the final states of the system; the δ-function presents the energy conservation law; E_i, E_f are the energies of the initial and final states of the system.

To define the wave functions $|i\rangle$, $|f\rangle$ and the disturbance operator, let us consider the potential where particle *A* moves during the reaction. The barrier parameters crossed by the particle are determined by the interaction of the reagents and potentials of the individual molecules shown in Figure 5.2a and b. It is obvious that if the reagents are far apart the potential energy of particle *A* over *R*—*A* and *A*—*R′* corresponds to Figure 5.2a and b. When the reagents come closer and, in particular, when they are at equilibrium, the mutual influence of the terms arises resulting in their transformation (Figure 5.3). Term 1—2 corresponds to potential energy *R*—*A* (Figure 5.2a), while term 1′—2′ corresponds to coupling *A*—*R′* (Figure 5.2b).

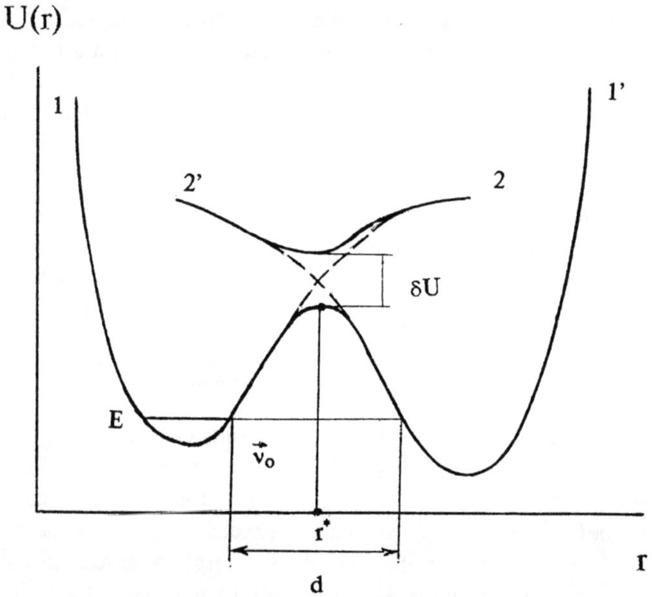

Figure 5.3 Intersection and quasi-intersection of terms, corresponding to Equation (5.1). Term 1-2 corresponds to Figure 5.2A and term 1′-2′ corresponds to Figure 5.2B. Here, d is the width of potential barrier, v_0 is the particle velocity under the barrier, and *E* is the particle energy.

The behavior of transition is correlated with the type of the potential curves or, more exactly, with the time required for readjustment of the electron subsystem. If in time (t^*), during which the particle is retained under the potential barrier,

$$t^* \cong d/|v_0|; \quad v_0 = \sqrt{2\left[E - U(r^*)\right]/m}, \tag{5.4}$$

the system fails to rearrange itself, the particle with term 1—2 tunnels over term 1′–2′ (Figure 5.3), and the transition is nonadiabatic (designations in Equation (5.4) are given in the caption to Figure 5.3 and in Equation (5.2a)). In another case when the particle motion under the barrier is slow, the electronic subsystem of the reagents manages to rearrange, the potential energy corresponds to term 1′—2′, and the tunneling is adiabatic.

It is natural that the type of transition also depends on the magnitude of the term splitting δU which is equal to the double exchange integral. Under a large splitting of the terms, the tunneling of the particle is adiabatic and at a small one when, for example, the symmetry or the spin of the system changes it is nonadiabatic.

5.1.4 Time of Transfer

Particle tunneling is appreciably influenced by the relation between the time used directly for the transfer and the period of parameter oscillations of a potential barrier. The role of these oscillations was briefly discussed in Sections 4.1 and 4.2 and will be treated in detail in this chapter. One should distinguish three basic events. First, the particle transfer can proceed faster (the process of the transfer is meant, not the waiting time of the transfer) than the intermolecular vibrations. In other words, it is as if the particle is waiting until the potential wells come together, and then at the minimum spacing between reagents, the transition under the fixed position of the surrounding molecules takes place. The second important case corresponds to the slow transfer of the particle at fast intermolecular vibrations. The transfer process takes place in the averaged potential. It is essential here that during the transfer the particle does not affect the parameters of the potential barrier. The transfer and the intermolecular vibrations occur independently. And, finally, the case can be realized when each position of the tunneling particle corresponds to its own configuration of the molecular environment determining the potential barrier of chemical reaction. Here there is a possibility for both the synchronous movement of the environment after the particle and the delay of the latter.

It has been shown that the possibility for transfer is influenced by the product of frequency of the intermolecular vibrations on the time of the tunneling transfer, Equation (5.4).[7] An account of such effects is required only in the case when this product is of one order of unity or over. If the width of the potential barrier is 1 Å and the height is 1.5 eV, then at a frequency of the intermolecular vibrations of 10^{12} to 10^{13} sec^{-1} for the transfer of hydrogen atom (these parameters are characteristic for radical reactions with the transfer of hydrogen atom[8-12]), this product makes up 10^{-1} to 10^{-2}, i.e., much less the unity. At the same time, for the reactions of heavy particles when this product is over unity, the overbarrier processes begin to play the main part (see Section 6.2.3).

5.1.5 The Double Adiabatic Approximation

It was already mentioned above that to obtain the numerical value of the rate constant it is necessary to know the wave function of the electron-nuclear system studied. Meanwhile the number of electrons and atoms participating in even a simple chemical reaction is rather large. This makes it practically impossible to solve this task correctly. In the given case, use should be made of the currently developed approximation techniques, in particular the double adiabatic approximation.

As is known, in considering a chemical reaction one can distinguish three types of subsystems: one electronic and two nuclear (inter- and intramolecular). The frequencies relating to the motions of these subsystems essentially differ. This is due to the magnitudes of the masses of the tunneling electrons and atoms (as a rule, H atoms) and molecules, as well as the potentials in which these aforesaid particles propagate. All this allows the wave function of the system to be written in the double adiabatic approximation by analogy with the conventional adiabatic approximation as the product of three functions corresponding to the electronic, inter-, and intramolecular subsystems.

It is necessary to note that in the matrix element from Equation (5.3), the integration by the coordinates of the electronic and intramolecular subsystems has already been carried out. As for the summations by the states of these subsystems, account is taken only of the initial state and the closest (from the bottom) final state of the electronic oscillating subsystem by the energy (Figure 5.1). Such an approximation is justified at not very high temperatures when the spacing to the closest level is essentially larger than the thermal energy. So the quantum energy of the C–H vibration is 0.3 eV which corresponds to a temperature of about 3500 K. The temperature up to which the sample where the reaction takes place remains solid is usually 300 to 400 K. Therefore, one can neglect the transitions from the higher states in all range of temperatures as long as the sample remains solid, since the population of these states is exponentially small.

This section gives a brief description of the main approximations required in construction of solid-phase chemical reactions theory. These approximations are also used in the theory of radiationless transitions, which is presented in References 13 and 14.

5.2 RATE CONSTANT OF SOLID-PHASE CHEMICAL REACTION

The rate constant of a radiationless transition in the solid phase is described by Equation (5.3). It should be noted, as mentioned above, that summations over the initial and final states should be performed only for the intermolecular subsystem. The argument of the δ-function has the form:

$$E_f - E_i = E_{\nu_f} - \Delta E;\ \Delta E = E_{n_i} - E_{n_f}.$$

Here, E_{n_i,n_f} is the energy sum of the electron and intramolecular subsystems for the initial and final states (Figure 5.1); E_{ν_i,ν_f} is the energy of the initial and final states of the intermolecular (slow) subsystem; and ν_i, ν_f are the quantum numbers corresponding to intermolecular vibrations. As a result, for the rate constant we obtain the expression:

$$k = \frac{2\pi}{\hbar}\sum_{v_i v_f} \exp\left[-\beta E_{v_i}\right] \left| < v_f \middle| V(R) \middle| v_i > \right|^2 \delta\left(E_{v_f} - E_{v_i} - \Delta E\right) \Big/ \sum_{v_i} \exp\left[-\beta E_{v_i}\right],$$

$$\beta = 1/k_B T. \qquad (5.5)$$

It is easy to see that to obtain the analytical formula in which all summations and integrations are represented it is necessary to have the means for tackling several questions. First, is it a type of transition matrix element type *V*(*R*) where *R* is the intermolecular coordinate? Second, in what approximation should we consider intermolecular vibrations? Finally, for what parameters of the medium and reagents should changes resulting from transition be noted?

Let us begin the discussion of these question with a type of matrix element. As is known, the calculation of sums similar to Equation (5.5) were performed using the theory of radiationless transitions. In doing so, various methods were employed: operator calculus,[15] method of generating polynomials,[16] and method of density matrix.[17] But, in all these papers summations were performed under the assumption that the matrix element *V*(*R*) is a constant which does not depend on intermolecular coordinates. This approach corresponds to the Franck-Condon approximation. In addition, works are available which account for the weak dependence of the matrix element on the intermolecular coordinates;[18,19] it expands in a power series in terms of reagent displacements from an equilibrium position. This approximation holds true for a transfer of both energy and electrons. Yet, in the case of chemical reactions when an atomic transfer takes place, a usual expansion cannot be used since a more precise account of the dependence of the matrix element on intermolecular coordinates is required. The transition matrix element is highly sensitive to a change of spacing between reagents. For example, to transfer a hydrogen atom through the characteristic potential barrier that is 1.5 eV high and 1 Å wide, zero intermolecular vibrations affecting the mutual arrangement and, especially, the spacing between reagents, can increase the matrix element by more than one order of magnitude.

It appears most reasonable to select an exponential series of matrix element dependence of atomic particle transfer on intermolecular coordinates. This is indicated by both the calculation of the overlapping integral for wave functions consistent with two parabolic or Morsian terms and the quasi-classical calculations of tunnel probability amplitude.[20-22] Consequently, we shall proceed from the ratio (see also Equation 4.8):

$$V(R) = V(R_0)\exp\left[-\frac{1}{2}J'(R_0)(R-R_0) - \frac{1}{4}J''(R_0)(R-R_0)^2\right],$$

$$J'(R_0) = -2\partial/\partial R \ln\left[V(R)\right]\Big|_{R=R_0}, \quad J''(R_0) = -2\partial^2/\partial R^2 \ln\left[V(R)\right]\Big|_{R=R_0}. \qquad (5.6)$$

Note that for a transition between two parabolic terms (Figure 5.4a) and in the parabolic potential (Figure 5.4b) this formula is exact. Only for functions $J'(R_0)$ and $J''(R_0)$ should one use corresponding expressions (see captions for Figure 5.4a and b); R_0 is the equilibrium values of the intermolecular coordinate.

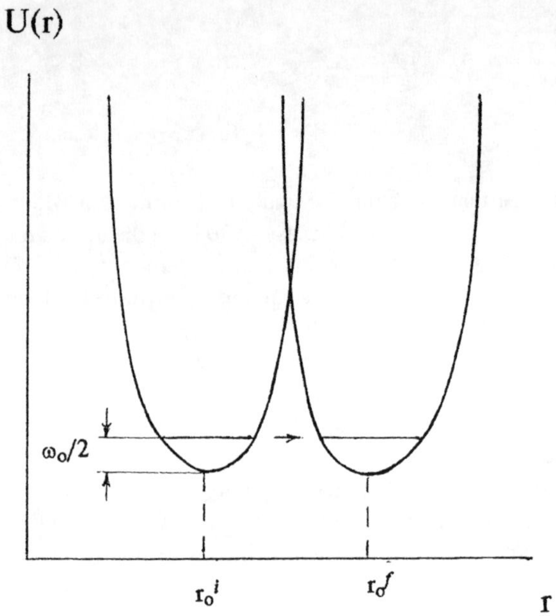

(A)

Figure 5.4 (A) Intersectual parabolic terms: parameters from Equation (5.5) are equal; $J' = m\omega_0 l_0/\hbar$, $J'' = m\omega_0/\hbar$, $l_0 = r_0^f - r_0^i$.

Thus, the necessity of presenting a matrix element in the form of Equation (5.6) does not allow the standard formulas of radiationless transfer theory to be applied for calculation of the rate constant. Along with this, it is precisely the account of the strong exponential dependence of the transition matrix element on reagent displacements from an equilibrium position that allowed the patterns of low-temperature solid-phase chemical reactions to be interpreted. For this, a modified theory of radiationless transitions[1,23] (in a special case of $V(R)$ = const leads to results reported in References 15 to 17). We note also that the developed apparatus of the modified theory of radiationless transitions is very close in its concept to the method of density matrix[17] and quantum-mechanical calculations of spectral lines pattern.[14]

Now we consider the description of intermolecular vibrations. In Chapter 4 it has already been mentioned that these vibrations are very important. It is the vibrations that provide for the approach of reagents which leads to a temperature dependence of the rate constant of solid-phase tunneling. In describing intermolecular vibrations we shall employ a harmonic approximation. On the one hand, this will allow all summations and integrations in Equation (5.6) to be performed analytically, and, on the other hand, such an approximation in a rather wide range of temperatures gives a satisfactory description of real molecular motions in solid phase. Only at relatively high temperatures close to the region of matrix "defrosting" when the

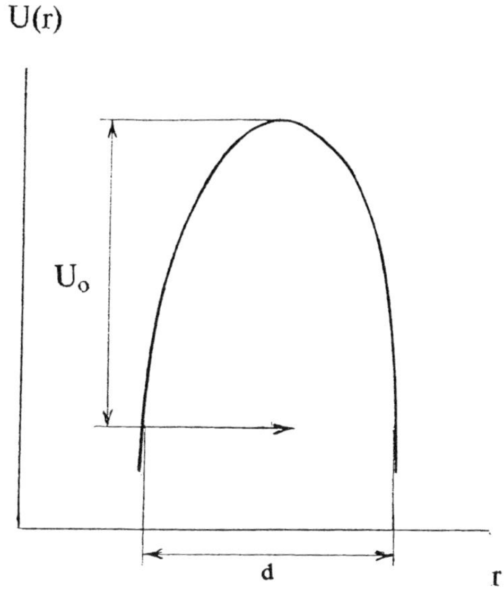

(B)

Figure 5.4 (B) Parabolic barrier: parameters from Equation (5.5) are equal; $J' = \frac{2\pi}{\hbar}\sqrt{2mU_0}, J'' = 2\pi\sqrt{2mU_0}/d\hbar$.

amplitudes of intermolecular vibrations increase does the anharmonieity start to play an appreciable role. This can be of particular importance in the range of temperatures close to a region of phase transition. Indeed, as is known, in the region of phase transition, the frequency of intermolecular vibrations can become rather small (soft mode). A decrease in frequency corresponds to an increase in amplitude of intermolecular vibrations and, respectively, the rate constant of tunneling increases.

Let us introduce the density matrix of the intermolecular subsystem:

$$\rho_\lambda(R,R') = \sum_\nu |\nu > \exp[-\lambda E_\nu] < \nu|.$$

Using the integral presentation of the δ-function from Equation (5.5) and separating in this expression the combinations formally corresponding to functions $\rho_\alpha(R, R')$, write down the rate constant by use of spurs:

$$k = \frac{1}{h}\int_{-\infty}^{\infty} dt \exp[-i\Delta Et] Sp\left\{V(R)\rho^i_{\beta+it}(R,R')V(R')\rho^f_{-it}(R,R')\right\} \Big/ Sp\left\{\rho^i_\beta(R,R')\right\}. \quad (5.7)$$

Here, the indexes i, f denote the density matrixes relating to the initial and the final states of the system.

To carry out the summations and integrations in Equation (5.7), the fact was used that all co-factors were exponential. This follows from Equation (5.6) and Reference 24, which contains the expression for the density matrix of the one-dimensional harmonic oscillator. In Equation (5.7) the density matrix of the multi-dimensional harmonic oscillator used, which represents the product of one-dimensional density matrixes and accounts for all normal vibrations of the lattice, is shown. A decisive factor for obtaining the analytical expressions is that the exponent presents the square form.

The suggested approach allows the rate constant formula of any radiationless process to be obtained when the equilibrium coordinates, the directions of the main axes, and the frequencies of the normal oscillations are different for the initial and the final states. Since the expressions in the braces from Equation (5.7) are exponential in the index which is the square form, the spur is calculated analytically although the final formula is rather cumbersome. Here,we shall not produce this expression entirely; its detailed derivation is provided in Reference 1 and only considered for some very important special cases. It is namely these cases that have been realized experimentally.[8-12]

Below we shall discuss in detail the solid-phase radical reactions. For this type of chemical reaction, a number of simplifications holds true, allowing for fairly simple expressions for the rate constant to be attained and comparison of these expressions with the literature experimental data on the hydrogen atom transfer to be carried out.

One basic simplification is that no account will be taken of the changes resulting from the transition of the system of normal coordinates and the frequencies of intermolecular oscillations. In addition, the magnitude and temperature dependence of the rate constant are influenced by such parameters of the reaction system as the energy defect of the fast subsystem, rearrangement of the medium, and reagents and intermolecular oscillations. In Section 5.3, we shall not account for the change of the equilibrium coordinates of medium molecules (rearrangement of the medium and reagents) since this parameter plays a part only at the charge transfer in the polar matrix and for the reactions with a great conformation rearrangement of molecules. Account will be taken only of the other two factors having importance in the solid-phase radical reactions.

5.3 RADICAL REACTIONS

Let us consider various approximations used in obtaining the rate constant of the solid-phase chemical reaction. All these approximations deal with the description technique of the intermolecular oscillations, because it determines the method of summations in Equation (5.7). Such a consideration makes it possible to obtain the expression for the rate constant as the function of temperature, reagent-to-reagent distance, and frequency of intermolecular oscillations or the phonon spectrum of the system. Here, we shall present the Einstein and Debye models of the medium and discuss the roles of the local mode and orientational motions.

5.3.1 The Einstein Model

The problem of finding expressions for the rate constant dependencies on medium and reagent properties and the temperature consists of calculating the sums and the integral in Equation (5.7). Since the transition probability is determined by the reagent-to-reagent spacing, it is necessary to describe the reciprocal motion of reagents near the equilibrium position. In the section that describes molecule vibrations, we shall use the Einstein model of the medium. In this case, it should be assumed that each molecule (including reagents) moves independently of the others, the motion being performed in the harmonic potential. The potential is created by the interaction with the environment, and the values of displacement of the lattice node (molecules) from the equilibrium position are the normal coordinates.

To make calculations in Equation (5.7) it is required to define the law of dispersion and state densities. In the Einstein model these values correspond to similar values for the one-dimensional harmonic oscillator. In accordance to the procedure suggested in Reference 1, we write down the formula for the atom transfer rate constant:

$$
\begin{aligned}
k &= \frac{2\pi}{\hbar}\left|V(R_0)\right|^2 \varphi_3^{-1/2} \exp\left[\Delta E/2k_BT + \varphi_1(1-\varphi_2)\right] I_{\Delta E/\hbar\Omega}(\varphi_1\varphi_2)\rho_f, \\
\varphi_1 &= \frac{(J')^2}{4}\left[\tanh(\hbar\Omega/4k_BT) + J''/2\right]^{-1}, \\
\varphi_2 &= \frac{\sinh(\hbar\Omega/4k_BT)}{2\cosh^3(\hbar\Omega/4k_BT)\left[\tanh(\hbar\Omega/4k_BT) + J''/2\right]}, \\
\varphi_3 &= \left[1+\frac{J''}{2}\coth(\hbar\Omega/4k_BT)\right]\left[1+\frac{J''}{2}\tanh(\hbar\Omega/4k_BT)\right], \\
J' &= J'(\hbar/M\Omega)^{1/2}, \quad J'' = J''(\hbar/M\Omega), \quad \rho_f = \hbar\Omega.
\end{aligned}
\tag{5.8}
$$

Here, Ω is the Einstein frequency, $I_{\Delta E/\hbar\Omega}$ is the Bessel function of imaging argument, ρ_f is the density of final state, and M is the medium molecules mass.

The use of Einstein's model for the description of solid lattice vibrations holds true only at relatively high temperatures when the motion of each node of the solid lattice can be considered independent. If these motions are classical when the arguments of the hyperbolic functions exceed the unity, then Equation (5.8) is simplified and assumes the form similar to Equation (4.9) where the force constant γ should equal $M\Omega^2/2$. In the temperature range limited from above by inequality $T < M\Omega^2/2J''k_B$, but as above for the classical molecular vibrations ($T > \hbar\Omega/4k_B$), the logarithm of the rate constant is proportional to temperature, see Equation (4.7).

With further temperature increase and when inequality $T > M\Omega^2/2J''k_B$ is met, the dependence becomes the Arrhenius one:

$$
k \sim \exp[-E_a/k_B\,T],\ E_a = (\mathrm{M}\Omega^2/4)(J'/J'')^2. \tag{5.9a}
$$

To determine the nature of the Arrhenius dependence and the physical meaning of the activation energy E_a we consider two parabolic terms (Figure 5.4a):

$$
U_i = \frac{\gamma_0}{2}\left(r - r_0^i\right)^2, \quad U_f = \frac{\gamma_0}{2}\left(r - r_0^f\right)^2, \quad \gamma_0 = m\omega_0^2.
$$

These terms mark the initial and the final states of the tunneling particle between which the transition takes place. The dependence of functions J' and J'' on the system parameters is presented in the caption of Figure 5.4a. If these parameter dependencies are substituted in Equation (5.9a), then the following ratio holds true for the activation energy:

$$E_a = \frac{\gamma_0}{2} l_0^2. \tag{5.9b}$$

The physical meaning of this relationship is very simple — the activation energy is equal to the energy used for the initial and final potential wells to come close till the total disappearance of the barrier.

5.3.2 The Debye Model

The assumption on which the Einstein model is based, that the molecules making up a solid oscillate independently, in many cases turns out to be insufficient. First of all, it applies to the region of low temperatures where the correlation of the lattice molecule vibrations is very important.

It is obvious that in case of the true three-dimensional lattice of the solid, even the ideal one, obtaining the analytical expression for the rate constant is an unsolvable task. For this reason we shall use the Debye model of the medium which is reliable in the region of low temperatures. The main approximations used in this model are as follows: first, for all acoustic modes (and at low temperatures only acoustic phonons play a part) the phonon frequencies coincide with and are proportional to the absolute value of the wave vector and, second, the Brilloin real zone is replaced by a spherical one. The Debye frequency ω_D is selected in such a way that the value of the wave number corresponding to it equals the value of the real and model zones in the wave vector space.

In these approximations and starting from Equation (5.7), in the region of low temperatures (the parameter J'' can be assumed to equal zero), we obtain the following for the transfer rate constant:

$$k \sim \exp\left[\left(\Delta E - |\Delta E|\right)/2k_B T + a'T^4\right],$$
$$a' = \frac{\pi}{60} \frac{(J')^2 k_B^4}{\hbar^3 \omega_D^5 M}. \tag{5.10}$$

In Equation (5.10) there are two items, the first of which depends on whether the energy is released or absorbed in the transition process.

If the energy is released ($\Delta E > 0$), this item is equal to zero; in the case of energy absorption ($\Delta E < 0$), the dependence becomes the Arrhenius one. We note that the finite low temperature limit of the rate constant is valid only for nonendothermic reactions. As far as the second item is concerned, the fourth degree exponent at temperature is the precise result which does not depend on the approximations of the model describing intermolecular vibrations. Near the bottom of the phonon zone — and this part of the spectrum is of most importance at low temperatures — the degree exponent is determined only by the density states. Only the coefficient α' is calculated on the Debye model basis.

At higher temperatures ($T > \hbar\omega_D/4k_B$) when intermolecular vibrations become classical as in the Einstein model, the rate constant is described by Equation (4.7). Further, if account is taken of the square term in the function $J(R)$ expansion, then the transition rate constant obtains the form which, under a further increase in temperature, is approaching the activational Equation (4.9).

Thus, only at low temperatures are there differences in the Debye and Einstein models (compare with Figure 4.5) of the rate constant description. When the temperatures increase the differences disappear and Equation (5.8) can be used for the rate constant in both models.

5.3.3 The Local Mode

In irradiated matter, as is known, the reaction-capable particles — radicals, ions, captured electrons — are induced. Their properties (mass, force constants of interaction with adjacent molecules, intramolecular spectrum) differ from the environmental molecules, which lead to a change of the spectrum and vibrational dynamics of the solid lattice. Such changes affect only a small region around the defect,[25,26] but it is in this region that the chemical transformations take place. That is, the regularities of the elementary act of the chemical reaction are essentially influenced by the fact that at least one of the reagents is a defect. Consequently, the matrix deficiency plays a role even in those cases when the total number of radiation defects (i.e., radicals) is small.

Earlier, primary attention was paid to studying the spectral characteristics of the phonon spectrum of the disordered solid, since they determine the dissipation processes as well as the thermodynamic and kinetic properties of the matter.[25] At the same time, for chemical processes, along with the spectral characteristics, of no less importance are the intermolecular motions of reagents, a fact which leads to the necessity for studying the vibrational dynamics of the disordered lattice.

Such a consideration was undertaken for the one-dimensional disordered lattice.[1,23,27] The formula is obtained for the atomic transition rate constant with note taken that one of the reagents (radical, ion) noticeably differs by mass and magnitude of interaction with the environment from the molecules composing the lattice. The analytical expressions were found for three special cases. Two of them are very interesting methodologically — when either the mass or the constant of interaction with one neighbour differs from the rest. The third most important case for solid-phase radiation chemistry meets the following experimental conditions: the mass of one of the reagents is less than the mass of the medium molecules and the bond constant of the defect with one neighbour is essentially larger than with the others. In this situation, the local mode plays the main role in the relative displacements of the reagents. The contribution of the vibrations corresponding to the undisturbed part of the phonon spectrum, as if displaced from the reaction zone, is proportional to the square of the constants ratio of the molecules bound in the regular part of the lattice and reagents with each other. A creation of the reagent-composed complex takes place, and the atomic transfer rate constant is determined by the internal vibrations of this complex. The rate constant Equation (5.8) holds true where Ω is the internal vibration frequency of the reaction complex, and M is its reduced mass.

We note that the exceptions of the atomic transfer rate constant relating to the solid lattice deficiency are very clearly developed in the low temperature region. Indeed, when the local mode plays the main part in the reagents approach (also for the Einstein model) a distinct low temperature plateau is observed (see Figure 4.5, the curve a). This happens due to the fact that at ($T < \hbar\Omega/4k_B$) the hyperbolic functions from Equation (5.8) cease to be dependent on the temperature, i.e., the intermolecular vibrations providing the reagents approach "become frozen". At the same time, for the ideal crystal at low temperatures, Equation (5.10) holds true, resulting in a smooth way out of the rate constant of the nonendothermal reaction to the low temperature limit. The reason for such behavior of the rate constant is that in the ideal crystal the phonon frequencies are distributed over the interval from 0 to ω_D, and even at rather low temperatures the number of phonons is not equal to zero. On the other hand, not all phonons contribute equally to the process of the reagents' approach. The low frequency phonons lead to the syn-phase motions of the molecules, and only the vibrations corresponding to the high-energy phonons cause the motion of the phonons in the anti-phase. It is mainly these phonons that determine the temperature role in the process of the reagents approach. Under a decrease in temperature, the quantity of the high-energy phonons continuously decreases which brings about the smooth temperature dependence (Figure 4.5, curve b) of the rate constant.

In Section 5.5, where the theory and experimental data are compared, it is shown how the matrix properties influence the regularities of the tunneling solid-phase radical reactions.

5.3.4 Orientational Oscillations

In the previous sections we demonstrated the important role of the translational intermolecular vibrations in forming the temperature dependence of the rate constant of solid-phase chemical reactions. In the region of temperatures $T < \hbar\omega_D/4k_B$, the reagents' approach takes place owing to zero intermolecular vibrations. The amplitudes of these vibrations which reduce the magnitude of the potential tunneling reaction barrier depend on the Debye frequency (or the frequency of the local mode). In particular, the contribution from zero oscillations to the exponent is inversely proportional to the Debye frequency.

One should bear in mind that in the solid phase, apart from the translational oscillations, molecules participate in the orientational oscillations, too. These motions can influence the parameters of the process of overcoming the potential barrier by the particle in the reaction process. Also, since the orientational oscillations have lower frequencies in comparison to the translational ones, they can both increase the contribution of zero frequencies and decrease the "defrosting" temperature of intermolecular motions. Besides, the dispersion ratios for the frequencies of the orientational and translational oscillations are different and also can lead to exceptions in the rate constant temperature dependence.

To clarify the role of the orientational intermolecular vibrations in the solid-phase particle transfer, the task of tunneling in the two-well potential whose parameters depend on the orientational phonon coordinates was considered.[1,28,29] Main attention was paid to the one-dimensional model, since this model assumes the analytical solution and retains all major regularities of the rate constant temperature dependence. In certain cases, the three-dimensional attribute of the solid-phase lattice is found essential; these aspects are mentioned separately.

The analysis of the role of orientational vibrations showed that there are considerable differences from what the translational vibrations yield. In the low temperatures region, the influence of the translational and orientational vibrations on the transition rate constant differs qualitatively with an increase in temperature when the intermolecular vibrations can be assumed classical, quantitatively. Thus, at low temperatures the translational vibrations (the Debye model) give the degree dependence of the rate constant logarithm on temperature (Equation (5.10)), and the orientational vibrations give the exponential. This is due to the fact that the orientational phonon zone starts not with zero but with the frequency $\omega_{min} > 0$[30] and for $T < \hbar\omega_{min}/4k_B$:

$$k \sim \exp\{A + B\exp[-\hbar\omega_{min}/k_B T]\}.$$

Note that such a type of temperature dependence takes place for the translational phonons in the Einstein model and for the local mode, but ω_{min} is noticeably less than Ω. It is known[31] that ω_{min} conventionally is in the interval from one to some tenths of centimeters^{-1}, respectively; the contribution to the exponent from the orientational vibrations at low temperatures in Ω/ω_{min} is larger than for translational vibrations. In the intermediate region, if $T > \hbar\omega_{min}/k_B$, the vibrations are still quantum and there are differences in the description of the one-dimensional and three-dimensional events. The temperature-dependent part of the exponent in the rate constant is proportional to $T^{1/2}$, and in the three-dimensional, $T^{3/2}$. In the range when the intermolecular vibrations are classical and the particle transition is tunneling, the inclination slope of the straight line describing the rate constant logarithm dependence on temperature for the orientational phonons is by an order of magnitude higher than the corresponding value for the translational phonons.

In the high temperature region, under the activation dependence of the transition rate constant on temperature, some differences from the translational vibrations also exist. For the orientational phonons the reason for the activation dependence is contained in the limited interval of the angles promoting the reagents' approach (the activation energy in this case is greatly determined by the geometry of the system), while in the case of translational phonons this is the combination of the potential wells for the barrier transfer of the particle. Situations are also possible when the turn of the molecules is necessary to fulfill the steristic conditions. This can happen, for instance, if the atom performing the transition in the course of the reaction is blocked by other atoms of the reagents or the environment. Hence, the pre-exponential factor will be dependent on the matrix elasticity and the steristic conditions that we usually are faced with in chemical kinetics, as well as the activation energy.

5.4 REACTIONS WITH THE IONS TRANSFER

In the radiation-chemical transformations, apart from the radical reactions, the ion-molecular processes play an important part. In the first works on the transfer of protons (see, e.g., References 32 and 33) and in the tasks relating to the transfer of electrons, account was taken only of the rearrangement of the medium. Later, the approach[1-3] taking into account the influence of intermolecular vibrations on the potential barrier parameters was extended to the charge transfer processes.[34-40] Accounts of the dynamic properties of the medium revealed their importance not only in radiation-chemical reactions but also in electrochemistry[40] and in biological processes, as well.[41]

It is evident that many problems of describing tunneling reactions with H-atoms transfer (see Sections 5.1 and 5.2) turn out to be important in the transfer of the charged particle, too. In particular, the relation between such values as the constant relaxation, transition time, frequency of intermolecular oscillations, and the rate constant of reactions decisively influence the dependence of these characteristics on temperature and the characteristics of the medium. In just the same way, the methods of vibrational spectroscopy and inelastic neutron scattering were used to study the proton transfer along the hydrogen bond in amides and peptides.[41]

The resolution time of these methods is 10^{-9} sec, which substantially exceeds the time characterizing the dynamics of coherent tunneling. This led to the necessity for improvement of measurement technique which allowed conclusions about the mechanism of the proton transfer to be drawn. In the given case it turned out that due to the high frequency of proton oscillations between the potential wells, the role of the modulations of the potential barrier parameters is decreased by the intermolecular vibrations. The proton readily oscillates between the potential wells at any spacing between heavy particles (N and O). It is easy to see that this process differs from the H-atom transfer in the chemical reactions where the expectancy time of the transfer appreciably exceeds the reserve frequency of intermolecular vibrations.

Thus, the problem of accounting for the intermolecular vibrations in the transfer process of the atomic charged particle is drawing the attention of researchers today. Also, in such processes in comparison to the radical reactions, a no less important task is classification of the medium rearrangement. That is, it is necessary to obtain the expression for the process rate constant accounting of all factors: medium rearrangement, changes of the potential barrier under intermolecular vibrations, and energy defect of the fast subsystem. Some regularities of the charge transfer, other than those that were explained in the above-cited literature, were considered earlier.[23,42-44] We dwell upon some of them.

At low temperatures and with the same assumptions used to obtain Equation (5.10) but taking into account the medium rearrangement in the exponent describing the rate constant of the charge transfer, an additional negative term appears. Its magnitude depends on the tunneling change of the phonon subsystem equilibrium coordinates. This term, especially in the polar media under a temperature tending to zero, can be rather large. Thus the probability for observation of the low temperature rate limit of the ion-molecular reactions in the polar medium is close to zero. Taking into account the role of medium repolarization, one can draw a conclusion about the possibility for a change of the proton transfer mechanism in the solid phase as the temperature decreases.

If, in the region of high temperatures, reactions such as:

$$RH^{+} + RH \rightarrow R + H^{+} RH,$$

take place as a result of the proton transfer, then at low temperatures in the polar matrix the mechanism can change. Due to a small probability of the medium rearrangement accompanying the charge transfer, the reaction would rather proceed through the H-atom transfer:

$$RH + RH^{+} \rightarrow R + HRH^{+}.$$

The reaction products in both cases are equal.

At an increase in temperature, the quadratic term in Equation (5.6) starts playing a substantial role. It can be accounted for within the Einstein model of the medium. The formula for the rate constant by its structure is rather similar to Equation (5.8). One should, however, introduce small changes in it. It is necessary to add the terms:

1. In the argument of the Bessel function, the term:

$$E_r/\hbar\Omega \sinh(\hbar\Omega/2k_B T), \text{ and}$$

2. In the exponent the function:

$$-E_r \coth(\hbar\Omega/2k_B T).$$

Here, E_r is the energy of the medium rearrangement.

This formula allows various limiting events to be investigated. For instance, in the high-temperature region the rate constant has the activation form. The activation energy is equal to the sum of two items. The first of them is determined by the medium rearrangement and the energy defect of the intramolecular subsystem, and the second one by the energy used for merger of the potential wells between which the transition takes place.

The analysis carried-out on transfer regularities was conducted without accounting for the possible disturbances of the ideal lattice introduced by the difference of the charge-medium interaction from the interaction of medium molecules with each other. If these differences are accounted for, then two events can be realized: first, when the charged particle interacts with all neighbours equally, and, second, when the charge particle creates a complex with one of the molecules of the nearest environment. The first event is simulated by a linear chain with the disorder in site 0 (Figure 5.5). This disorder is characterized by the fact that the elastic constant of the sites' –1.0 and 0.1 differs from the bond constant between the rest of the sites of the lattice. After the charge passes from site 0 in site 1, the change of the normal coordinates system takes place. Therefore, the transition rate constant in the arbitrary system of the coordinates was found not to coincide with either the initial or the final. In the second instance, which is also simulated by a linear chain with the disorder in the knot 0, only the bond constant between sites 0 and 1 differs from the rest (Figure 5.6).

Expressions for the transition rate constant accounting for a defect influence were obtained.[44] The consideration showed that due to the distortion of the phonon spectrum and the appearance of local modes in the medium, both the absolute value of the rate constant and its temperature dependence change. In this case, the absolute value is greatly determined by the local mode and the temperature dependence in

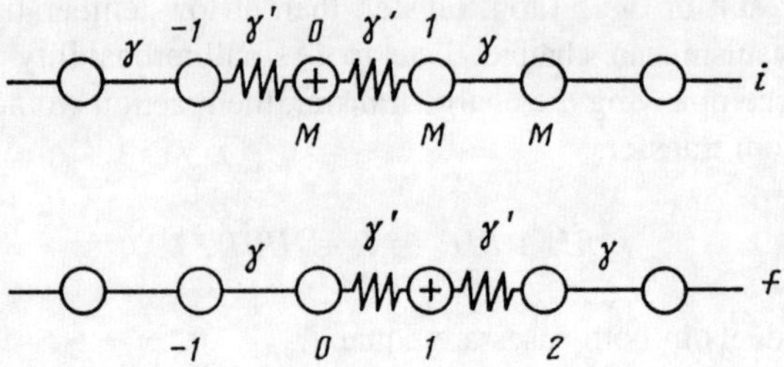

Figure 5.5 One-dimensional model of charge transfer in polar media. Transfer is accompanied by change of system normal coordinate. Upper chain (*i*) corresponds to the initial and lower chain (*f*), to the final state of the system.

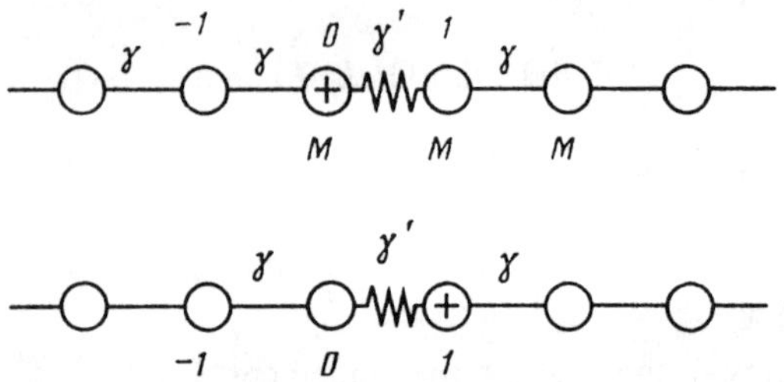

Figure 5.6 One-dimensional model of charge transfer in polar media. Transfer occurs without changes of system normal coordinate. Upper chain (*i*) corresponds to the initial and lower chain, (*f*) to the final state of the system.

the region of low temperatures — by vibrations relating to the dense part of the phonon spectrum.

5.5 COMPARISON OF THEORETICAL AND EXPERIMENTAL DATA

Let us compare the literature of experimental data relating to the H-atom transfer in radical reactions (see Section 4.3.3) with the theory described in this section.

Four reaction systems are considered: methyl radical with molecules of methanol[8] and ethanol[10] alcohol, H-atom transfer in γ-radiated dimethylglyoxime,[9] and creation of radical couples in the fluorene-acridine system.[11,12] In the first two systems, the local mode plays the main role in the approach of reagents under intermolecular vibrations. Correspondingly, in treating the experimental data, we shall use Equation (5.8), since the temperature in these experiments is relatively small, the parameter J'' can be assumed equal to zero.

As for the processes in dimethylglyoxime and in the mixture of fluorene-acridine, they are crystals and comparison of theoretical with experimental is conducted somewhat differently. At low temperatures, when $T < \hbar\omega_D/2k_B$, Equation (5.10) should be used to describe the experimental data. As a consequence of meeting the condition $\Delta E \geq 0$ the activation term in Equation (5.10) is equal to zero.

With the increase in temperature, the rate constant dependencies for the Debye and Einstein (and also for the local mode) models coincide (Figure 4.5). Therefore, a simpler expression can be employed for the Einstein model, Equation (5.8). It should be kept in mind that the experiments in the fluorene-acridine system were carried out in a very wide range of temperatures — up to 300 K — and for the natural transition to the activation dependency it is necessary to take into consideration the square term in Equation (4.8). Equations (5.8) and (5.10) for obtaining temperature dependence over all investigated ranges of temperatures match under the same values of parameters. Furthermore, we shall assume that the Debye and Einstein frequencies are equal in value.

Comparison of theory and experiments was carried out in the following way. First of all, the potential surfaces were constructed. For the reactions that were studied in References 8 to 10, the LEPS method[45] and the collinear model of the chemical reaction were utilized; for the transfer of H atoms in the fluorene-acridine system,[11,12] the semi-empirical method AMI[46] was used. Further, using the constructed potential surfaces the function, $J(R)$ is calculated:

$$J(R)=\frac{2}{\hbar}\int_{a_1(R)}^{\alpha_2(R)} dx\left\{2m\left(U(x,R)-E(R)\right)\right\}^{1/2}.$$

Here, it is taken into account that the energy ($E(R)$) and the potential energy ($U(x,R)$) of the tunneling particle are the functions of the molecular coordinate R. It defines the boundaries of potential barrier a_1 (R) and a_2 (R). Utilizing the function $J(R)$, it is easy to calculate its derivatives $J'(R_0)$ and $J''(R_0)$ that determine the temperature dependencies of Equations (5.8) and (5.10).

The technique of fitting the experimental and theoretical data for reactions[8-10] is described in detail in Reference 1. Below, a comparison is carried out and discussion of the results is undertaken for all four reactions.

Figures 5.7 and 5.8 show the results of adjustment for the reactions of H-atom transfer from alcohol molecules to the methyl radical. The optimum values of parameters at which the agreement of theoretical and experimental data is the best are given in the captions to the figures. First of all, we note that both the distances and the intermolecular frequencies for ethanol and methanol happened to be very close in value. It is no wonder, since the systems studied are of the same type. In addition, the frequency values obtained somewhat exceed the Debye frequencies in such systems. This agrees with the properties of the local mode whose frequency must exceed the Debye frequencies (Section 5.3.3).

The processing of experimental data on the transformation of radical pairs in the γ-irradiated crystal of dimethylglyoxime gives the following values of the adjusting parameters: R_0 = 3.17 Å and $\omega_D = 3.10^{13}$ sec^{-1} (Figure 5.9).[9] The change of the molecular structure as a result of irradiation is small and a comparison can be made of the designed value of the equilibrium spacing between the oxygen atoms (R_0) with values 3.1 to 3.2 Å known from the literature.[47] It is easy to see that the agreement is very good.

The equilibrium spacings in dimethylglyoxime proved to be less than analogous spacings in alcohol matrices. The difference value is in the range of 0.4 to 0.5 Å and makes up about 0.5 Å. The difference of the equilibrium spacings can be estimated using the values of the van der Waals radii and bond lengths:

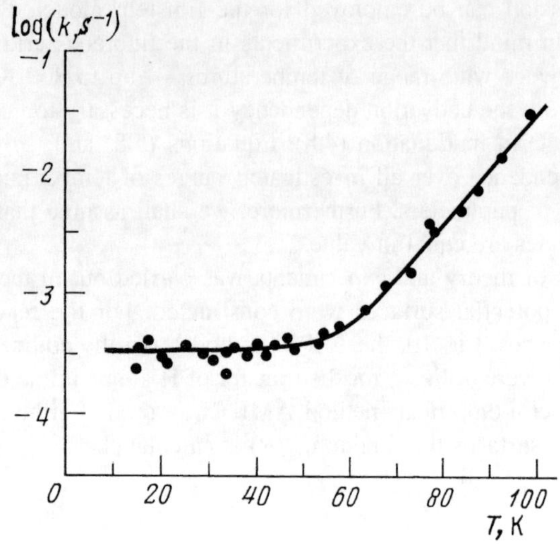

Figure 5.7 Temperature dependence of rate constant of hydrogen-atom transfer from the methanol molecule to methyl radical. • = experiment;[8] — = fitting curve. Values of the fitting parameters are equal to R_0 = 3.66 Å, Ω = 3.56 · 10^{13} sec^{-1}.

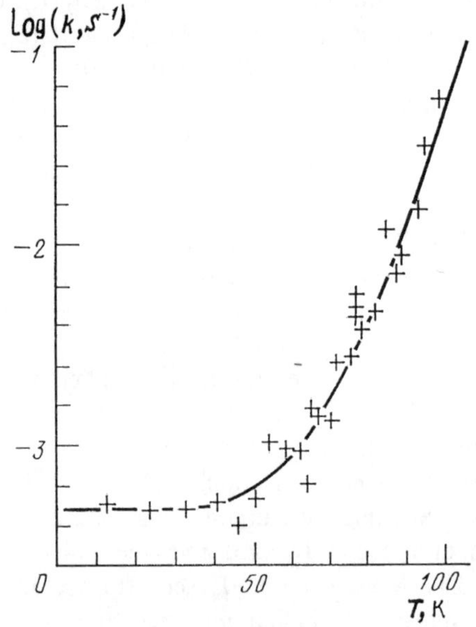

Figure 5.8 Temperature dependence of rate constant of hydrogen atom transfer from the ethanol molecule to methyl radical. + = experiment;[10] — = fitting curve. Values of the fitting parameters are equal to R_0 = 3.64 Å, Ω = 3.5 · 10^{13} sec^{-1}.

$$R_0(\text{C–H---C}) - R_0(\text{O–H---O}) =$$
$$(\rho^{\text{C–H}} + \rho^{\text{H}} + \rho^{\text{C}}) - (\rho^{\text{O–H}} + \rho^{\text{H}} + \rho^{\text{O}}) =$$
$$(1{,}1 + 1.17 + 1{,}8)\text{Å} - (0.96 + 1{,}17 + 1.52)\text{Å} \approx 0.4\ \text{Å}.$$

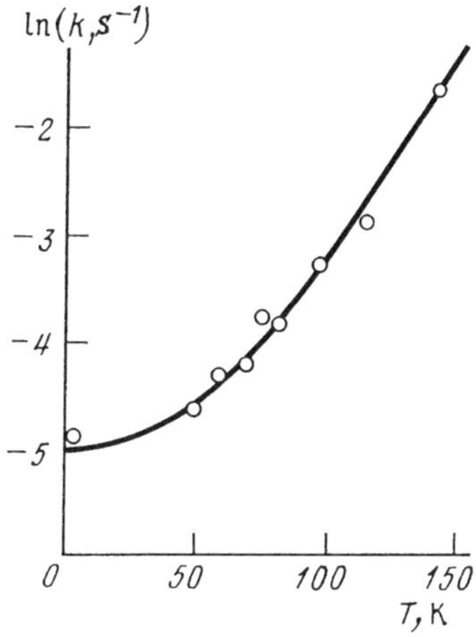

Figure 5.9 Temperature dependence of rate constant of hydrogen atom transfer in irradiated crystal of dimethylglyoxime. O = experiment;[9] — = fitting curve. Values of the fitting parameters are equal to R_0 = 3.17 Å, ω_D = 2.92 · 10^{13} sec^{-1}.

It is easy to see that the expected difference and the one obtained by means of matching the difference value of the equilibrium distances are in good agreement.

Now, we turn to interpretation of the experimental data on the H-atom transfer from fluorene to acridine.[11,12] To spare machine time, the potential surfaces were calculated for the quasi-molecule consisting of molecules of acridine and cyclopentadine simulating fluorene.[48,49] The molecules are located in positions corresponding to the crystallograph (Figure 5.10). The spacing between the molecules' planes was 3.22 Å.

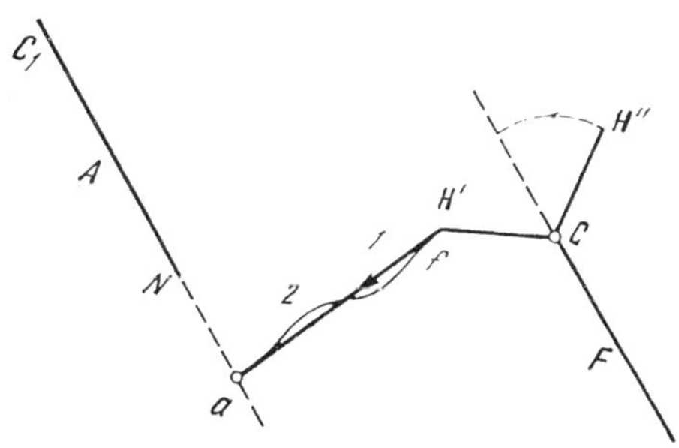

Figure 5.10 Scheme of hydrogen atom transfer reaction in fluorene-acridine system. The explanation of the figure are given in the text.

The possibility for such a modeling was confirmed by direct calculations of the potential curves for the hydrogen atoms transfer in the fluorene-acridine system in the ground and triplet states. The maximum difference in comparison with the model calculation was less than 0.04 eV in the potential barrier maximum region. The path of the H-atom transfer reaction from the fluorene molecule (F) to the acridine molecule (A) is shown in Figure 5.10. The H′ atom moves straightforward (straight line 1) from position f to position a. Simultaneously, the H″ atom moves along the arc of the circle to the plane of molecule F. The calculations show that the minimum energy path has a form different from the straightforward (curve 2). However, these deviations are small and a one-dimensional path of the reaction can be considered.

Figure 5.11 shows the potential curves for the process in the ground (curve 1) and triplet (curves 2, 3, 4, 5) states with different interplane spacings. These results allow the parameters J′ and J″ to be calculated. Then the calculated rate constant is compared with the experimental data (Figure 5.12). The best agreement is achieved under R_0 = 3.24 Å and $\omega_D = 2.14 \cdot 10^{13}$ sec^{-1}. For the low temperature limit, the value k_0 = 0.23 sec^{-1} is obtained that satisfies Equation (4.11). The magnitude obtained for R_0 is somewhat less (by 0.13 Å) than the crystallographic value. This is quite possible, because the atomic transition follows the electron excitation and distorts the acridine molecule. In addition, the real position of the acridine molecule in the crystal lattice of fluorene is unknown.

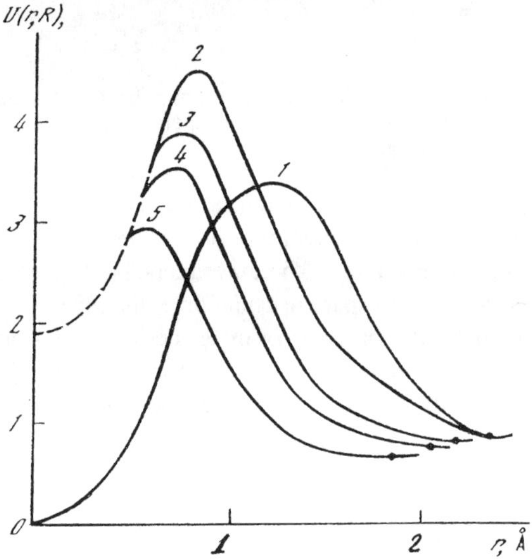

Figure 5.11 Potential energy curves for the process of a hydrogen atom transfer in the singlet and triplet states of the fluorene-acridine system. The cases of different distances between the molecules' planes are considered: 1, 2 are the singlet and triplet transfers at crystallographic distances between planes; 3, 4, 5 are the triplet transfers when the distance between planes is less than crystallographic on 0.2 Å, 0.3 Å, 0.489 Å, respectively. At these distances, the heights of the barrier are 2.6 eV (curve 2), 1.98 eV (3), 1.68 eV (4), and 1.04 (5). The other values of parameters are given in the text.

The values R_0 can be compared with the experimental data in dimethylglyoxime (see above). The difference is equal to 0.24 Å. We estimate this difference proceeding from the van der Waals radii and the lengths of the corresponding bonds:

$$R_0(\text{C–H---N}) - R_0(\text{O–H---O}) =$$
$$(\rho^{\text{C–H}} + \rho^{\text{H}} + \rho^{\text{n}}) - (\rho^{\text{O–H}} + \rho^{\text{H}} + \rho^{\text{O}}) =$$
$$(1.1 + 1.17 + 1.58)\text{Å} - (0.96 + 1.17 + 1.52)\text{Å} \approx 0.2\ \text{Å}$$

that corresponds to the above value.

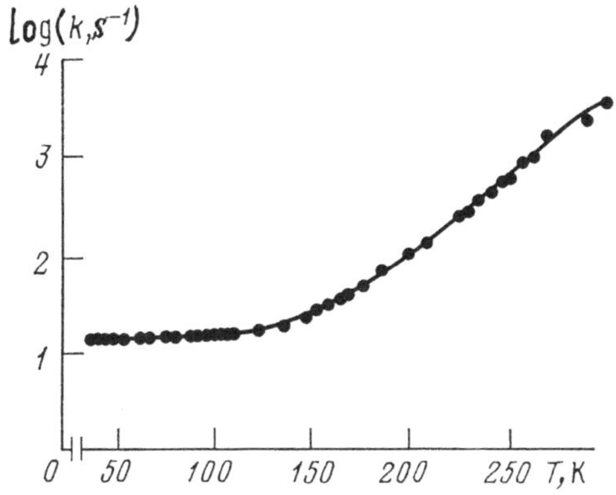

Figure 5.12 Temperatuere dependence of rate constant of hydrogen atom transfer in the fluorene-acridine system. • = experiment;[11,12] — = fitting curve. Values of the fitting parameters are equal to R_0 = 3.42 Å, $\omega_D = 2.14 \cdot 10^{13}$ sec^{-1}.

The obtained value ω_D is close to the literature data on the molecular crystals with rigid bonds: $\omega_D = (1.5 - 3.5) \cdot 10^{13}$ sec^{-1}. In the fluorene-acridine system, the frequency $2.5 \cdot 10^{13}$ sec^{-1} becomes apparent and totally corresponds to the value calculated from the experimental data.

There are two more works in which the theoretical and experimental data were compared with respect to the dependence of the H-atom transfer rate constant on temperature in the fluorene-acridine system.[50,51] Under rather precise calculations of the potential surface, the use of the one-mode intermolecular vibrational model did not allow the regularities of the process to be shown, especially in the low-temperature region. The reason for this is in accordance with the experimental conditions, refusal from the Debye model of the fluorene crystal lattice. The analysis of different approaches to the interpretation of experimental data[11,12] is given in Reference 52.

Apart from studying the temperature dependence of the atomic transfer rate constant, a number of publications are available in which further evidence of tunneling has been suggested. The question concerns the kinetic isotope effect, which is greatly dependent on the mass of a particle performing transition.[6] The most convincing demonstration of the tunneling mechanism of solid-phase reactions was obtained in Reference 53. At the temperature of 77 K, the kinetic isotope effect was studied in the reaction of H-atom abstraction by means of a methyl radical from the molecule of acetonitril. It exceeded by one order of magnitude the maximum possible value of classical isotope effect. The value measured: $\eta = k_H/k_D \geq 2.8 \cdot 10^4$, was later compared with the calculations[54] that yielded the value $\eta = 5 \cdot 10^5$. This corresponds to the experimental data.[53] We note that a similar value ($\eta = 10^{3.3}$ to $10^{5.4}$) has been obtained earlier on the basis of the classical S-matrix method.[55]

REFERENCES

1. Goldanskii, V.I., Trakhtenberg, L.I., Fleurov, V.N., *Tunneling Phenomena in Chemical Physics,* Gordon and Breach Science, New York, 1989.
2. Hanggi, P., Talker, P., Borkovek, M., Reaction-rate theory: fifty years after Kramers, *Rev. Mod. Phys.,* 62, 251, 1990.

3. Cleland, A.N., Martinis, J.M., Clark, J., Measurement of the effect of moderate dissipation on macroscopic quantum tunneling, *J. Phys. Rev.*, B37, 5950, 1988.
4. Devoret, M.H., Esteve, D., Martinis, J.M., Urbina, C., Turlot, E., Effect of an adjustable admittance on the macroscopic energy levels of a current biased Josephson junction, *Phys. Scr.*, 39, 121, 1989.
5. Benderskii, V.A., Goldanskii, V.I., Tunneling of heavy particles in the low temperature chemistry, *Int. Rev. Phys. Chem.*, 11, 1, 1992.
6. Landau, L.D., Lifshitz, E.M., *Quantum Mechanics*, Elmsford, New York, 1981.
7. Ivanov, G.K., Kozhushner, M.A., Nonadiabatic effects in tunneling reactions of heavy particles, *Sov. J. Chem. Phys.*, 2, 1299, 1983.
8. Hudson, R.L., Shotany, M., Williams, F., Hydrogen atom abstraction by methyl radicals in methanol glasses at 15–100 K. Evidence for a limiting rate constant below 40 K by quantum mechanical tunneling, *Chem. Phys. Lett.*, 48, 193, 1977.
9. Toriyama, K., Nunome, K., Iwasaki, M., Evidence for tunneling hydrogen atom transfer reactions at 4.2 K in organic crystals, *J. Amer. Chem. Soc.*, 99, 5823, 1977.
10. Le Roy, R.J., Murai, H., Williams, F., Tunneling model for hydrogen abstraction reactions in low-temperatures solids. Applications to reactions in alcohol glasses and acetonitril crystals, *J. Amer. Chem. Soc.*, 102, 2325, 1980.
11. Prass, B., Colpa, J.P., Stehlik, D., Identification of the lowest energy nuclear fluctuation modes promoting the photochemical H-transfer tunneling reaction in doped fluorene single crystals, *J. Chem. Phys.*, 88, 191, 1988.
12. Prass, B., Colpa, J.P., Stehlik, D., Intermolecular H-tunneling in solid state photoreactions promoted by distinct low-energy nuclear fluctuation modes, *Chem. Phys.*, 136, 187, 1989.
13. Kun, H., Rhys, A., Theory of light absorption on non-radiative transition in F-centres, *Proc. Roy. Soc.*, 204, 406, 1950.
14. Lax, M., The Franck-Condon principle and its application to crystals, *J. Chem. Phys.*, 20, 1752, 1952.
15. Feyman, R., An operator calculus having application in quantum electrodynamics, *Phys. Rev.*, 84, 108, 1951.
16. Pekar, S.I., About effect of lattice deformation by electrons on optical and electrical properties of crystals, *Russ. Chem. Rev.*, 50, 197, 1953.
17. Kubo, R., Toyozawa, Y., Application of the method of generating function to radiative and nonradiative transitions of a trapped electron in a crystal, *Progr. Theor. Phys.*, 13, 160, 1955.
18. Bikbayev, N.Kh., Ivanov, A.I., Lomakin, G.S., Ponomarev, O.A., Theory of radiationless processes in "non-Condon" approximation, *Sov. Phys. JETF*, 74, 2154, 1978.
19. Levine, A.M., Hontscha, W., Pollak, E., Sudden theory for tunneling in dissipative systems, *Phys. Rev.*, B40, 2138, 1989.
20. Perlin, Yu.E., Modern methods of multiphonon processes theory, *Russ. Chem. Rev.*, 80, 553, 1963.
21. Siebrand, W., Radiationless transitions in polyatomic molecules. I. Calculation of Frank-Condon factors, *J. Chem. Phys.*, 46, 440, 1967.
22. Siebrand, W., Radiationless transitions in polyatomic molecules. II. Triplet-ground-state transitions in aromatic hydrocarbons, *J. Chem. Phys.*, 47, 2411, 1967.
23. Trakhtenberg, L.I., Klochikhin, V.L., Pshezhetskii, S.Ya., The theory of the tunnel transition of atoms in solids, *Chem. Phys.*, 69, 121, 1982.
24. Husimi, K., Some formal properties of the density matrix, *Proc. Phys.-Math. Soc. Jap.*, 22, 406, 1940.
25. Maradudin, A.A., *Theoretical and Experimental Aspects of the Effects of Point Defects and Disorder on the Vibration of Crystals*, Academic Press, New York, 1966.

26. Ziman, J.M., *Principles of the Theory of Solids,* Cambridge University Press, London, 1972.
27. Trakhtenberg, L.I., Effect of intermolecular vibrations on the rate of tunneling reactions in solids, *Sov. J. Chem. Phys.,* 1, 53, 1982.
28. Trakhtenberg, L.I., Shteinschneider, N.Ya., Role of orientational vibrations in tunneling reactions in solid phase, *Russ. J. Phys. Chem.,* 60, 1409, 1989.
29. Trakhtenberg, L.I., Flomenblit, V.Sh., Effect of orientational intermolecular vibrations on the rate hydrogen transfer from fluorene to acridine, *Russ. J. Phys. Chem.,* 67, 1464, 1993.
30. Anselm, A.I., Porfireva, N.N., Orientational-translational waves in molecular crystals. I. Dynamics of linear chain, *Sov. Phys. JETF,* 19, 438, 1949.
31. Alecsandrov. I.V., Zhidomirov, G.M., Calculation of spin-lattice relaxation time for radicals in molecular crystals, *Sov. Phys. JETF,* 41, 127, 1961.
32. Bell, R.P., *The Proton in Chemistry,* 2nd ed., Cornell University Press, New York, 1973.
33. Ulstrup, J., *Charge Transfer Processes in Condensed Media,* Springer-Verlag, New York, 1979.
34. Borgis, D.C., Lee, S., Hynes, J.T., A dynamical theory of nonadiabate proton and hydrogen atom transfer reaction rate in solution, *Chem. Phys. Lett.,* 162, 19, 1989.
35. Borgis, D.C., Hynes, J.T., Molecular-dynamics simulation for a model nonadiabate proton transfer reaction in solution, *J. Chem. Phys.,* 94, 3619, 1991.
36. Shida, N., Almloff, J., Barbara, P.F., Tunneling paths in intramolecular proton transfer, *J. Phys. Chem.,* 95, 10457, 1991.
37. Special issue of *J. Phys. Chem.,* 95, 25, 1991.
38. Sokolov, N.D., Vener, M.V., Proton tunneling assisted by the intermolecular vibration excitation in solid state, *Chem. Phys.,* 168, 29, 1992.
39. Smedarchina, Z., Enchev, V., Lavchieva, I., Theoretical study of intramolecular proton transfer in solution: application to the photoenolization of 5,8-dimethyl-1-tetralone, *J. Phys. Chem.,* 98, 4218, 1994.
40. Cappadonia, M., Kornyshev, A.A., Krause, S., Kuznetsov, A.M., Stimming, U., Low temperature proton transport in clathrates, *J. Chem. Phys.,* 101, 7672, 1994.
41. Kearley, G.J., Fillaux, F., Baron, M.-H., Benington, S., Tomkinson, J., A new look at proton transfer dynamics along the hydrogen bonds in amides and peptides, *Science,* 264, 1285, 1994.
42. Trakhtenberg, L.I., Effect of medium on chemical reaction rate in solid phase, *Sov. J. Chem. Phys.,* 2, 1683, 1983.
43. Klochikhin, V.L., Trakhtenberg, L.I., Effect of intermolecular vibration on the charge transfer rate in solid phase, *Sov. J. Electrochem.,* 19, 1212, 1983.
44. Klochikhin, V.L., Trakhtenberg, L.I., Role of the crystal lattice defects in charge transfer at low temperatures, *Sov. J. Electrochem.,* 22, 1429, 1986.
45. Eyring, H., Lin, S.H., Lin, S.M., *Basic Chemical Kinetics,* John Wiley & Sons, New York, 1983.
46. Dewar, M.J.S., Zoebisch, E.G., Healy, E.F., Stewart, J.J.P., AMI: a new general purpose quantum mechanical molecular model, *J. Amer. Chem. Soc.,* 107, 3902, 1985.
47. Merit, L., Lauterman, E., The crystal structure of dimethylgloxime, *Acta Crystallogr.,* 5, 811, 1952.
48. Flomenblit, V.Sh., Mikheikin, I.D., Trakhtenberg, L.I., Cooperative tunneling transfer of hydrogen atoms in the fluorene-acridine system, *Dokl. Phys. Chem.,* 320, 695, 1991.
49. Flomenblit, V.Sh., Trakhtenberg, L.I., The temperature dependence of the rate constant of a hydrogen transfer in solid-state reaction within a fluorene-acridine system, *Sov. J. Chem. Phys.,* 11, 1931, 1994.

50. Chantranupong, L., Wildman, T.A., Golden-rule treatment of hydrogen abstraction by photoexcited acridine guests in fluorene single crystals, *J. Chem. Phys.,* 94, 1030, 1991.
51. Lavtchieva, L., Smedarchina, Z., Generalized golden rule treatment of photochemical solid-state reaction involving hydrogen tunneling, *Chem. Phys. Lett.,* 184, 545, 1991.
52. Chan, T.V., Wong, C.M., Stehlik, D., Pressure dependence of the low temperature tunneling rate for the hydrogen transfer in acridine-doped fluorene crystals, *Chem. Phys. Lett.,* 219, 187, 1994.
53. Sprague, E.D., Deuterium isotope effect on hydrogen atom abstraction by methyl radicals in acetonitrile at 77 K. Evidence for tunneling, *J. Chem. Phys.,* 81, 516, 1977.
54. Klochikhin, V.L., Trakhtenberg, L.I., Effect of intermolecular vibration on the rate and isotope effect of chemical reactions in solid phase, *Sov. J. Chem. Phys.,* 2, 810, 1983.
55. Ovchinnikova, M.Ya., The tunneling dynamics of low temperature H exchange reactions, *Chem. Phys.,* 36, 85, 1979.

6 KINETIC PECULIARITIES OF SOLID-PHASE RADIATION-CHEMICAL REACTIONS

Consideration of solid-phase processes taking place under irradiation shows that the kinetics of practically all stages of radiolysis are not entirely conventional. Therefore, this chapter sequentially discusses the stages of the radiation-chemical process that usually lead to creation, transformations, and annihilation of chemically active particles. The question, as a rule, will be about radicals, but many of the described regularities relate to charged particles, too.

It is known that under the effect of γ-radiation creation of radicals takes place. Apart from creation, this radiation brings about their annihilation. It has been shown that the correlation of the efficiency of these processes and the account of the migration of free valences under the effect of radiation determine such important characteristics of radiolysis as the limit concentration value of radicals, its independence on the absorbed dose rate, and the presence of the anomalously large radiation yield of recombination.

The mechanisms explaining the kinetics of radical reactions are discussed below. Both reactions with H-atom transfer (tunneling reactions) and heavy atomic particles participation are considered.

Possible reasons for the appearance of such radiation-chemical phenomena as the cage effect and the compensation effect are discussed. The decisive role of the dynamic properties of the medium is emphasized.

The final stage of substance radiolysis in the solid phase is the recombination of radicals. The peculiarities of this process are related to the restriction of radicals mobility in the solid phase.

All these problems are fundamental to solids radiation chemistry. There is constant interest in them, and one cannot say that there are simple answers to all questions related to the kinetics of radical processes. Some mechanisms of the mentioned phenomena have been clarified; in other cases, some difficulties due to interpretation of the results are observed. In addition, possible explanations of the observed patterns are suggested.

6.1 RADICAL ANNIHILATION IN THE PROCESS OF THEIR ACCUMULATION UNDER RADIATION

When solids are being irradiated, the concentration of radicals first increases proportionally to the dose, then the rate of their accumulation gradually decreases and the concentration attains the limit value (Figure 6.1). This limit concentration does not depend on the absorbed dose rate and is in the range of 10^{-3} to 10^{-5} from the initial molecule concentration of the irradiated medium. To explain these regularities, several mechanisms were suggested;[1-3] however, none of them satisfactorily describes the kinetics of radicals accumulation in the process of irradiation. In particular, the magnitude of the limit concentration and the anomalously large yield of the recombination are not explained.[4,5]

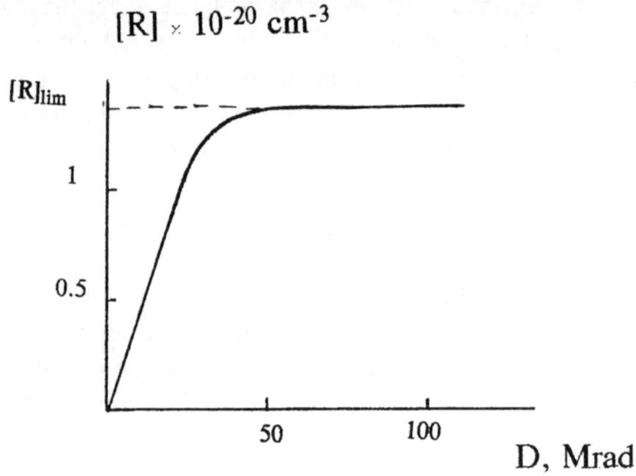

Figure 6.1 Characteristic dependence of the radical concentration from the absorbed dose. $[R_{lim}]$ is the limited radical concentration.

This chapter offers a mechanism of recombination of free radicals in the solid phase based on their absorption of excitons arising from irradiation. This leads to the dissociation of radicals. The produced mobile radical — as a rule, a hydrogen atom reacting with the environmental molecule — abstracts a hydrogen atom from it, producing a new radical and a molecule. Thus, a diffusion of free valences takes place until two radicals are so close to each other that they can recombine. This process is similar to the development of a photoradical chain (see Chapter 7). In this case, the exciton plays a part in the light quantum.

Let a solid be irradiated by ionizing radiation with the dose rate j; G is the radiation-chemical yield of radicals on the average in volume; G_1 is the yield of radicals outside the spurs; G_2 is the yield of radicals in the spurs; G_e is the yield of excitons.

The absorption of excitons leading to the dissociation of radicals, as is known, takes place with the rate

$$w = 4\pi\, D_e L N_e,$$

here, D_e is the diffusion coefficient of the excitons; L is the radius of exciton capture by the radical; N_e is the concentration of excitons. We shall take into account that the excited radicals dissociate with a probability close to the unit.[6]

The length of the diffusion path of the mobile radical (further, we shall assume that it is a hydrogen atom) until it enters into reaction with the molecule is equal to:

$$l = \sqrt{D_H \tau_1} = 1/N_m \sqrt{\sigma_H \sigma_1}.$$

Here, D_H is the diffusion coefficient equal to $v/(3\sigma_H N_m)$ where v is the velocity of the hydrogen atom; σ_H is the cross-section of its elastic collisions with substance molecules whose concentration is equal to N_m; τ_1^{-1} is the probability per unit time of the reaction $H + HR \rightarrow H_2 + R$; $\sigma_1 v\sigma_1$Nm is the cross-section of this reaction; $\tau_1^{-1} = v\sigma_1 N_m$.

The time of a free valences transfer is determined mainly by the rate of H-atom creation ($\tau_1^{-1} >> w$). Starting from this and making use of the expression for the absorption rate of the excitons and the length of the free path of the mobile radical, one can find out the diffusion coefficient of free valences (D) and the rate constant of free radicals recombination (k_r):

$$D = \frac{l^2 w}{3}, \qquad k_r = 4\pi D L_1.$$

Here, L_1 is the radius of the radicals recombination.

Now, let us consider the kinetics of creation of radicals under irradiation of a solid by ionizing radiation. Separately, we shall study processes in the spurs and in the whole volume. In the spur, the initial number of radicals n_0 in the volume V_0 is created and then a recombination of the radicals according to the described mechanism and a diffusion fuzzying of the spur take place. It is assumed that in the process of evolution in the spur the additional creation of radicals does not happen. The loss of radicals is described by the square-law. The dependence of the spur volume on time (t) is determined by the diffusion fuzzying of the spur. For a spherically symmetric spur the solution has the form[7]

$$m = n_0 \left[1 + \frac{2}{5} \frac{n_0 k_r}{CDV_0^{1/3}} \left(1 - \frac{1}{1 + CDtV_0^{-2/3}} \right) \right]^{-1}, \qquad (6.1)$$

$$C = 8(4\pi/3)^{2/3}.$$

As follows from Equation (6.1), upon the end of time $t > t_1 = V_0^{2/3} (CD)^{-1}$, the number of radicals in the spur aims for its stationary value independent of the absorbed dose rate. As for the volume-averaged radiation-chemical yield of radicals G, it is equal to $G_1 + G_2 \dfrac{n(t \rightarrow \infty)}{n_0}$. Utilizing the expression for G, it is possible to derive the equations describing the mean concentration of radicals and excitons.[7] In a general case, the formulae are very bulky. Therefore, let us consider limiting situations often encountered in experiment:

1. The concentration of radicals (R) is small and the absorption of excitons by radicals is not essential; then

$$R = R^a_{\lim}\tanh\left(k^a jt\right),$$
$$R^a_{\lim} = \frac{N_m}{4\pi}\sqrt{\frac{G}{G_e}\frac{3\sigma_1\sigma_H}{D_e\tau_e LL_1}}, \qquad k^a = \sqrt{\frac{G}{G_e}}\Big/R^a_{\lim}. \tag{6.2}$$

2. The lifetime of excitons is determined by their capture by radicals; then

$$R = R^b_{\lim}\left(1-\exp\left[-k^b jt\right]\right),$$
$$R^b_{\lim} = \frac{G}{G_e}\frac{3\sigma_1\sigma_H N_m^2}{4\pi L_1}, \qquad k^b = G/R^b_{\lim}. \tag{6.3}$$

We emphasize that in both limiting cases the value $R_{\lim}$ does not depend on the radiation dose that corresponds to the experimental data.[4,5,8] The experimental data on the accumulation of free radicals in the solid phase cited in these references are satisfactorily described by expressions of the type in Equations (6.2) and (6.3).

Now we carry out estimations for the case of creation of radicals under irradiation of cyclohexane. Let us assume $\sigma_H = 2\cdot10^{-15}$ cm^2, $V_0^{1/3} = 2$ nm, $L_1 = 0.4$ nm, $G_e = 4$, $G = 2$. The cross-section σ_1 of the reaction H + H$R \rightarrow H_2 + R$ noticeably depends on the H-atom energy; in our case, this energy is much larger than the thermal and $\sigma_1 \approx 10^{-16}$ cm^2. Indeed, the hydrogen atom is not able to become thermalized as a result of collisions: at the time of its creation it possesses the energy ΔE equal to the difference of the energies of the electron-excited state ($E_e < 4.2$ eV[9]) and the C–H bond breakdown energy in the radical ($E_b < 3.7$ eV). The loss of energy under elastic collision does not exceed the Debye temperature (θ_D = 100–200 K) and the characteristic time of thermalization is $t_{\text{therm}} \cong (\Delta E/\theta_D)\,(v\,\sigma_H N_m)^{-1}$. This value essentially exceeds the time necessary for the atom to enter into reaction $t_r = l/v = (v\sqrt{\sigma_H\sigma_1 N_m})^{-1}$. The ratio of the times is $t_{\text{therm}}/t_r = (\Delta E/\theta_D)\sqrt{\sigma_1\sigma_H} >> 1$.

Assuming $\sigma_1 \approx 10^{-16}$ cm^2, we derive from Equation (6.3) $R^b_{\lim} = 3\cdot10^{19}$ g^{-1}, $k_b = 6\cdot10^{-22}$ g/eV, which corresponds to the experimental data.[8] The energy yield of radicals recombination per 100 eV of the absorbed energy is $k_b N_m = 4\cdot10^2$. This corresponds to the results of References 4 and 5.

It was established that the kinetics of the free radicals accumulation in radiolysis is described by the ratio of Equation (6.3). The estimated values $k_b = (1.6 + 0.5)\cdot10^{-21}$ g/eV, $R^b_{\lim} = (1\text{–}6)\cdot10^{19}$ g^{-1} for proteins and polypeptids and $k_b = (6.6\cdot10^{-20}\text{–}6.6\cdot10^{-22})$ g/eV, $R^b_{\lim} = (3\cdot10^{17}\text{–}3\cdot10^{19})$ g^{-1} for a number of hydrocarbons are sufficient to correspond to the results of our estimations.

Earlier we believed that the fuzzying of the spur took place quickly. This is true, indeed. The estimations show that in cyclohexane the characteristic time of spur fuzzying under an irradiation rate of 20 kGy/h is of the order of 5 sec, i.e., much less than the time of irradiation.

In conclusion, we note that Equations (6.2) and (6.3) make it possible to estimate qualitatively the dependence of a radical limit concentration on temperature T. It is known that the radius of radicals recombination L_1 increases as the temperature is raised.[10] Furthermore, the diffusion coefficient of excitons depends on temperature. Thus, for noncoherent excitons $D_e \sim \exp(-E_a/k_BT)$ (E_a is the activation energy). Hence, when the lifetime of excitons is not determined by their absorption by radicals, $R^a_{\lim} \sim L_1(T)\exp(E_a/2k_BT)$. If the annihilation of excitons is determined by their capture by radicals, then $R^b_{\lim} \sim L_1(T)$, i.e., in both cases, the value $R_{\lim}$ drops with the T growth which corresponds to the experimental results.[10]

6.2 KINETICS OF RADICAL REACTIONS

In a number of experimental studies, it has been shown that the kinetics of chemical reactions in solid matrices do not obey the law of the first or the second order and has a more complicated form (see also Section 4.3):[11-13]

$$C(t) = C(0)\exp[-Kt^{\alpha}],\ 0 < \alpha \geq 1.$$

Here, $C(t)$ is the concentration of chemically active particles at time t. The parameter values α and K are determined by matrix properties and temperature. In particular, in a glassy alcohol matrices[12,14,15] the dependence of radical concentration on time is also described by Equation (6.4), with the values of α being changed in the range $\alpha = 0.26$ (30 K) to $\alpha = 0.77$ (100 K).

There are earlier studies[16,17] where α in the range of temperatures 77 to 87 K was assumed to be equal to 0.5. The experimental results[16,17] are explained in References 18 and 19 and more fully presented in Reference 20. Also, Reference 21 should be mentioned for its discussion of very similar mechanisms of nonexponential kinetics of low-temperature radical reactions.

Briefly, the results obtained in References 18 and 19 are as follows. It has been shown that the spatial nonuniformity of the matrix can lead to the nonstationary kinetics of solid-phase cage reactions.[16,17,22,24] Account is made of the exponential dependence of the tunneling reaction rate constant on the intermolecular distance. So, in decreasing the distance between reagents by 0.03 to 0.04 Å, the transfer rate constant increases by an order of magnitude.[20] Averaging by equilibrium intermolecular distances gives kinetics of radicals reactions with environmental molecules different from the first order law. The time dependence of the initial radicals concentration at 77 K is well described by Equation (6.4) with $\alpha = 1/2$ in the range of two orders of magnitude in depth of transformation.

Interpretation has been given to the kinetics of methyl radicals in the mixtures of normal and deuteriated alcohols at the temperature of liquid nitrogen.[18] The linear dependence of derivative methyl radicals concentration by the $\sqrt{t}$ law vs. the nondeuteriated molecules proportion in the matrix was observed. Such a dependence is observed if the distance between the methyl radical and molecules in the cage is not correlated.[16] This corresponds to the fact that the distances considered are independent arbitrary values, and their distributions for different cages are equal.

Experimental data was compared with theory for the kinetics of radical reactions in mixed alcohol matrices (C_2H_5OD, C_2D_2OD).[16] Reactions with three types of free radicals — CH_3, C_2H_5, C_4H_9 — were investigated. The evaluations showed that the number of molecules in the cage does not increase as the radical grows but rather sharply decreases.[19] The reason for this is due to the fact that in the process of "searching" a partner radical can leave the initial cage. Apart from this, it is the decreased mobility of the radical once its value is increased that leads to the decrease in "cage dimension".

Another mechanism is possible,[15,25] based on the matrix relaxation which results in the orientation of reagents suitable for a chemical process. In Section 6.2.2, these mechanisms are compared. In the conclusion of the section, the regularities of reactions with the transfer of heavy particles Cl and Br (see Section 4.3.4) are discussed. It is shown that account should be taken of the important role of the energy released as a result of the reaction and stored in the amorphous substance.

6.2.1 Polychronic Kinetics of the Tunneling Cagial Reactions in the Solid Phase

Let us consider the experimental data on the basis of Equation (6.4) within the framework of polychronic kinetics perception.[12,14] It is assumed that as the temperature is raised, a decrease of equilibrium states dispersion takes place since the growth of temperature removes the spatial restrictions in the mutual location of molecules.[26,27] Hence, the kinetics of the process loses the "braked" character that must lead to an increase in parameter α at the temperature growth. Naturally this parameter must strive for unity.

To consider the temperature dependence of the dispersion parameters of equilibrium distances, we restrict ourselves to a simple case of the polychronic model which is characterized by two parameters — the minimum (k_{min}) and maximum (k_{max}) values of the rate constant. We assume that the logarithms of these rate constants are distributed randomly with the interval

$$\delta \ln k = \ln(k_{max}/k_{min}). \tag{6.5}$$

Within the framework of the given model the dependence of radicals concentration on time has the form

$$c(t)/(0) = \left(\delta \ln \bar{k}\right)^{-1} \int_{k_{min}}^{k_{max}} (k)^{-1} \exp[-kt]dk =$$

$$Ei(x) = \int_{x}^{\infty} s^{-1} \exp(-s)ds = \left(\delta \ln \bar{k}\right)^{-1}\left[Ei\left(k_{min}t\right) - Ei\left(k_{max}t\right)\right]. \tag{6.6}$$

Using Equations (6.6) and (6.5) we calculate the kinetics of radicals reaction with the molecular environment at different widths of the distribution $\delta \ln k$ in the coordinates $\ln \ln[c(0)/c(t)] - \ln(k_{max}t)$ (see Figure 6.2). As a result of such a consideration it may be noted that there are rather wide sections of linear dependence which correspond to Equation (6.4). From the series of curves presented in Figure 6.2, we find a relation between the parameter α and the width of the logarithm distribution of the rate constant, (Equation (6.5)). The calculation results are given in Figure 6.3. Using this data (choosing values $\alpha(T)$), we produce the dependencies of radicals concentration on time calculated by Equation (6.6) in the coordinates corresponding to the empiric law, Equation (6.4); (see Figure 6.4). As a result of this calculation, it becomes clear that the dependence (6.4) can manifest the polychronic kinetics of solid-phase radical reactions.[12,14,18]

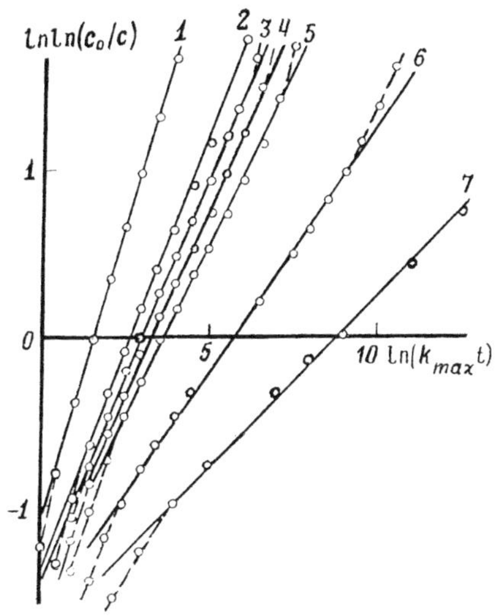

Figure 6.2 Interpolation of polychronic kinetics described by Equation (6.6) (points) with the dependence (6.4) (solid line). Width of logarithm distribution of rate constant, Equation (6.5) is equal to 3 (straight line 1), 5 (2), 5.5 (3), 6 (4), 6.67 (5), 10 (6), 20 (7).

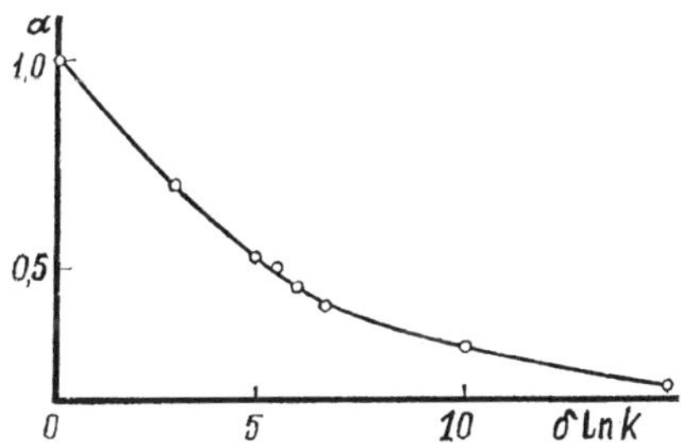

Figure 6.3 Dependence of exponent α of the function, Equation (6.4), from the distribution width δln*k* in the polychronic model.

The dependencies of ethyl radical concentrations on time behave in a similar way in the temperature range 77 to 107 K. The hydrogen atom transfer in the reaction of ethyl radical with a three-ethylamine molecule was investigated.[27] The parameter α changed in the range 0.5 to 0.75.

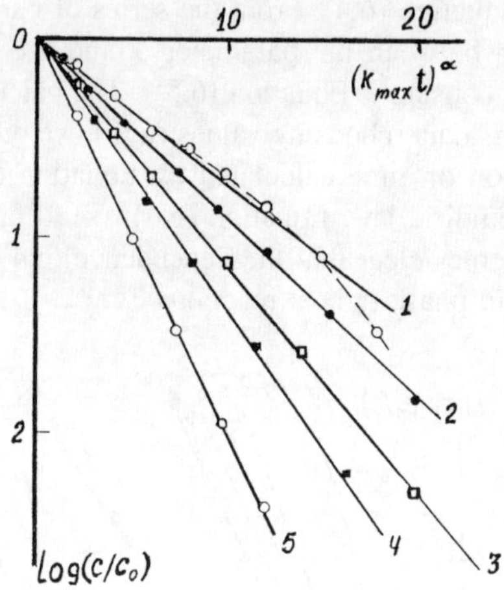

Figure 6.4 Polychronic kinetics, Equation (6.6), in coordinates which correspond to the dependence, Equation (6.4): 1–δln$k$ = 10 (α = 0.3); 2–5.5 (0.5); 3–4 (0.6); 4–3 (0.67); 5–1.9 (0.8).

We note that in the isotope-replaced matrices at the initial part of the kinetic curve the dependence on the portion of reactivity molecules initial concentration ($c(0)$) does not have the form $(c(0)t)^{\alpha}$, but $c(0)t^{\alpha}$. Exactly this dependence was observed in References 12 to 14.

Now we interpret experimental data[12,14] on the kinetics of low-temperature reactions of the H-atom abstraction by a methyl radical in glassy matrices of non-deuteriated alcohols on the basis of the polychronic model. Figure 6.5 shows experimental data and calculation results (solid line) of the kinetic curve by Equation (6.6).[12,14] This curve at short times cannot be described by the empiric ratio of Equation (6.4). With longer times, the kinetic curve describing the annihilation of the main quantity of radicals CH_3 obeys the dependence of Equation (6.4). It should be taken into consideration that each theoretical curve in Figure 6.5 is calculated at those values that fit parameters ln k_{min} and ln k_{max} to which the distribution functions, presented in Figure 6.6, correspond. It is obvious that the experimental data in the whole interval of times is satisfactorily described by the calculation curves.

The form of distribution in Figure 6.6 allows the explanation that the initial reaction rate at 20 K is to be even somewhat larger than at higher temperatures 30 and 40 K. At the same time the annihilation of the main quantity of radicals CH_3 at 20 K ocurs essentially slower than at higher temperatures. This evidently is due to the fact that at 20 K freezing of rather large and energy-consuming fluctuations of the liquid phase density initially contained at room temperature is possible. In the fluctuational regions with increased density the intermolecular distances are smaller. This essentially manifests on the H-atom tunneling rate. As mentioned above,[18] the distribution of the logarithms of the reaction rate constant correspond to a distribution of tunneling distances δR by formula $\delta \ln \bar{k} = J'\delta R$ (where $J' = 0.8 \cdot 10^{10}$ cm^{-1}). From here, for the widest distribution at 20 K, we arrive at the spread of widths $\delta R \cong 0.2$ Å, and for a width distribution $\delta \ln \bar{k} =$ 2.2 at 100 K we obtain $\delta R \cong 0.3$ Å.

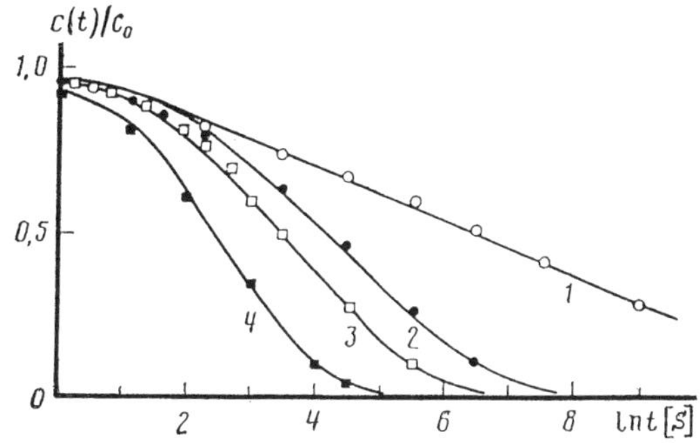

Figure 6.5 Interpolation of experimental data (points) on kinetics of low temperature reactions of hydrogen atom abstraction by methyl radical in glassy alcohol matrices.[12,14] Description of this process was based on the polychronic model, Equation (6.6) (curves) at the temperatures 20 K (curve 1), 77 K (2), 87 K (3), 100 K (4). Distributions of the rate constants given in Figure 6.6 correspond to the distributions of these curves.

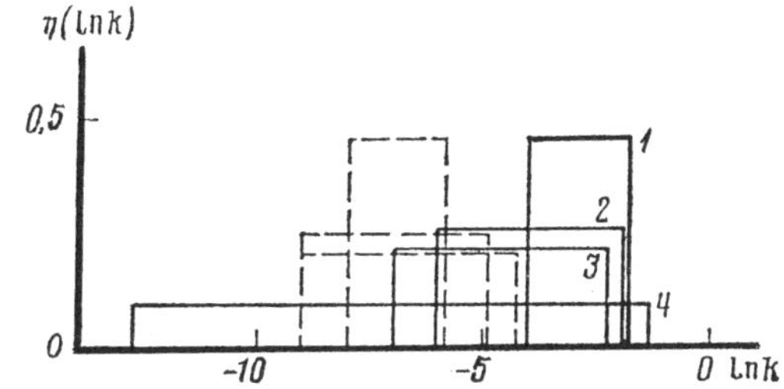

Figure 6.6 Distribution function of rate constant logarithm. These distributions correspond to the experimental kinetics of Figure 6.5. Distribution 1 corresponds to a temperature of 100 K; 2, 87 K; 3, 77 K; 4, 20 K. Distribution represented by the dashed line is explained in the text.

Now let us discuss possible reasons for temperature dependence of the logarithms of reaction rate constants. According to Figure 6.6, two trends of temperature influence on the distribution function can be separated: (1) with the increase in temperature, the distributions become narrow, and (2) with the increase in temperature, the gravitational centers shift towards larger reaction rates. As a result of the addition of two trends, the position of the right-hand edge of the distribution (maximum constants) does not depend on temperature which leads to a weak temperature dependence of the initial reaction rate at short times.[12,14]

According to the solid-phase reaction model,[20,28] with H-atom tunneling the temperature affects not only the equilibrium spacing between reagents but also the rate constant value directly. Due to this fact, the temperature dependence of the reaction elementary act leads to a shift of the whole distribution of the logarithms of reaction rate constants towards larger values without changing the distribution width. The observed restriction of distribution functions can be explained only by a change of the matrix properties: at higher temperature the matrix becomes better annealed and correspondingly more ordered which leads to a decrease in the spread of tunneling distances and rate constant logarithms.

The temperature dependence of the elementary act rate constant for a given reaction[20,28] shows that at 40 K the natural logarithm of the rate constant increases approximately by 0.5; at 77 K, by 2; at 87 K, by 3; at 100 K, by 4. Using these values we can exclude the temperature dependence of the reaction elementary act from the total dependence shown in Figure 6.6. As a result there remains a dotted temperature dependence stipulated by the degree of the matrix nonuniformity. To explain a form of distribution, i.e., to evaluate the spread of distances at which the hydrogen atoms tunnel, one should keep in mind that the unity in the scale $\ln\bar{k}$ corresponds approximately to 1/80 Å. It is obvious that a new distribution (dotted line) undergoes a sufficient narrowing with the temperature increase, but the shift does not take place. This can be interpreted in the following way: the gravitational center of this distribution corresponds to some "unstrained" configuration similar to all temperatures shifted neither towards the larger nor towards the smaller media densities.

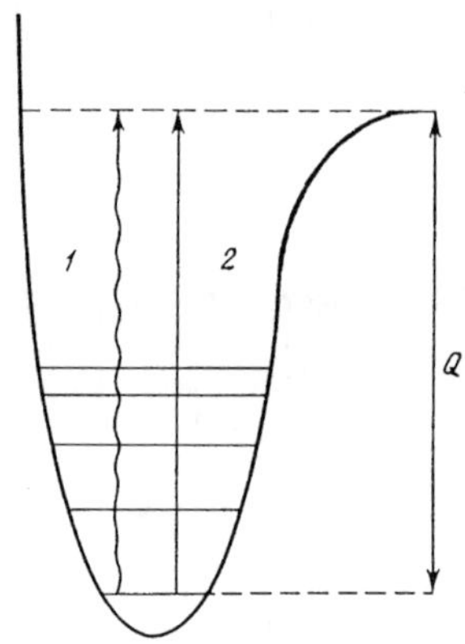

Figure 6.7 Scheme illustrating the particle motion in energy space. The boundary is reached because of thermal motion (1) and straight transition (2).

A comment should be made in relation to the expediency of presenting experimental data on reactions with polychronic kinetics in the form of dependencies of Equation (6.4). It is established that these dependencies are purely empirical and contain little physical meaning.[12,14,16] We show how discrepancies of the experiment, and in particular unprecise knowledge of the initial concentration $c(0)$, can distort the value of the empirical parameter α. Figure 6.4 illustrates that at the distribution widths $\delta \ln R = 3$ and 10 the polychronic kinetics straightens best of all in the coordinates $\ln (c/c(0)) - Kt^{\alpha}$ at $\alpha = 0.67$ and 0.3, respectively. Nevertheless, if an attempt is made to straighten it in the "universal" coordinates $\ln (c(t)/c(0)) - Kt^{1/2}$,[18] then good agreement is obtained if one neglects some small interval of the initial times.[20]

Hence, if the experiment does not allow the measurements to be carried out in the interval of short times and there is no confidence in the safety of determining the initial concentration $c(0)$, then fitting the degree dependence to the experimental data with the aim of finding the optimal value α is not justified. Taking this into account, the more natural coordinates for describing the polychronic kinetics are $c/c(0) - \ln t$ (see Figure 6.5).

In the conclusion, we note that the given fitting procedure ($\ln c/c(0) = K_\alpha t^\alpha$) is weakly sensitive to the changes of α in the interval 0.3 to 0.4 under these conditions, and certainly, that simultaneously, the value K_α is varied in the optimal way. This can be the reason (as mentioned in Reference 12) that the shape of the kinetic curves (or, its equivalent, the exponent) of abstraction by a methyl radical of hydrogen atom from α-position and from β-position in ethanol coincides. The type of the kinetics also does not change when the free radical C_2H_5 is replaced by C_4H_9. A consequence of this fact can be universality of the law $\ln((c(t)/c(0)) = -Kt^{1/2}$ by means of which a broad class of radical reactions in glassy materials is described successfully.[15,16] The value $\alpha = 1/2$ turns out to be successful for fitting the polychronic kinetics because it is intermediate between $\alpha = 1$ (usual kinetics) and $\alpha = 0$ (completely "broken" kinetics). Nevertheless, the values $\alpha = 0.4$ or 0.6 evidently would be equally successful as universal parameters.

One can agree with the authors[12] that for the radical reactions in matrices containing a deuteriated fraction of molecules the cagial polychronic kinetics lead to a number of molecules accessible to a radical sufficiently exceeding the coordination sphere.[19] Hence, a free radical leaves the boundaries of the coordination sphere. But this is not the reason for refusal from the polychronic kinetics if it embeds into the notion of the cage of not only the first but several successive coordination spheres.

In the most general formulation polychronic kinetics are the kinetics of a sum of independent kinetics for each realization of the environment of a free radical. And as long as the concentration of radicals in the matrix is sufficiently small each of them reacts with the environment independently, and the total kinetics is polychronic. The individual kinetics can include migration of radicals beyond the first coordination sphere and creation of reaction-capable configurations and an elementary act of H-atom abstraction. All these processes in total will determine the spectrum of lifetime or the kinetics of the annihilation of free radicals in the matrix.

6.2.2 Nonexponential Kinetics of Chemical Reactions in the Relaxation Matrix

It has been noted above that the nonexponentiality of kinetics can be caused by both the spread of equilibrium spacings between reagents and the relaxation of molecular environment. We consider these possibilities in more detail.

First, we take into account only the spread of the rate constant due to micro nonuniformity of the matrix when relaxation processes are disregarded completely. In this case, the dependence on the time of concentration of active particles has the form:

$$C(t) = C(0)\int_0^\infty g(k)\exp(-kt)dk, \tag{6.7}$$

where $g(k)$ is the unit-normalized function of the rate constant distribution.

According to the second standpoint, the matrix relaxation plays the defining role, the result of which is the nonstationary state of the kinetics. Here, it is assumed that the reaction happens only upon reaching a certain local state or a group of states by the microdomain of the reagents-contained matrix. Until such a state is reached, the reaction does not proceed, and, consequently, the relaxation is a limiting stage of the process. In this case, the dependence of reagents concentration on time often is described by formula

$$C(t) = C(0)\exp\left[-\int_0^t K(t')dt'\right], \tag{6.8}$$

where $K(t)$ is the time-dependent "constant" of the reaction rate.

In this work the matrix relaxation and the reaction are considered on an equal basis. It turns out that the dependence of reagents concentration on time always can be presented in one of the above-mentioned forms. We emphasize that these expressions are equivalent and the time-dependent rate constant $K(t)$ is connected with the function $g(t)$ by the relation to:

$$K(t) = \int_0^\infty kg(k)\exp(-kt)\,dk \Big/ \int_0^\infty g(k)\exp(-kt)dk. \tag{6.9}$$

In general, instead of the rate constant there can appear generalized parameters characterizing the time evolution of the relaxed matrix with simultaneously occurring reactions in it. The important conclusion of this consideration is the fact that the observance of delayed kinetics does not give an answer to the question about the reasons for its nonstationarity.

The influence of matrix relaxation on the kinetics in the presence of the rate constants spread is considered in Reference 29. However, this work disregarded the relaxation polychronity, i.e., it was assumed that the matrix is characterized solely by the time of relaxation. In this case, the nonstationary kinetics caused by the matrix relaxation turned out to be "out of sight". Now let us consider the above-discussed kinetics.[30] To do this, we renumber the local states of reagents and the matrices with index j = 1,2,...,N. The time dependence of the subsystems concentration $C_j(t)$ in the state j is determined by equation

$$\frac{dC_j(t)}{dt} = -\sum_{j' \neq j} w_{jj'}C_j(t) + \sum_{j' \neq j} w_{j'j}C_{j'}(t) - k_jC_j(t) = -\sum_{j'=1}^{N} T_{jj'}C_j(t). \tag{6.10}$$

Here, $w_{jj'}$ is the probability for the subsystem to transit from state j into state j' relative to the unit time; k_j is the chemical reaction rate constant for the subsystem in state j. The two first terms describe the matrix relaxation, while the latter term describes the chemical transformation of reagents in reaction products.

It is easy to see that the rate constant does not depend on time in two cases: either all partial rate constants k_j are equal or the matrix relaxation takes place substantially faster than the chemical transformation does. In the first case, $K = k_j$

= *const*. In the second case, quasi-stationary partial concentrations are established quickly.

Since Equation (6.10) is linear, its general solution is found easily:

$$C_j(t) = \sum_{i,j'} S_{ji} e^{\frac{-t}{\tau_i}} \left(S^{-1}\right)_{ij'} C_{j'}(0). \tag{6.11}$$

Here, S_{ij} is the matrix diagonalizing the matrix T_{ij} introduced in Equation (6.10), the proper values of which are equal to $1/\tau_i$, i.e.,

$$\left(S^{-1}TS\right)_{ij} = \frac{1}{\tau_i}\delta_{ij}.$$

Summing up the partial concentrations of Equation (6.11), we find the dependence of reagents concentration on time:

$$C(t) = \sum_{i=1}^{N}\left[\sum_{j=1}^{N}\left(S^{-1}\right)_{ij} C_j(0)\right]\left[\sum_{j=1}^{N} S_{jt}\right] e^{\frac{-t}{\tau_i}} \tag{6.12}$$

The time constants τ_j characterize the matrix relaxation and reaction happening simultaneously. The values $1/\tau_i$ and k_i are equal only when the matrix relaxation can be disregarded, i.e., $w_{jj'} = w_{j'j} = 0$.

In this case,

$$C(t) = \sum_{j=1}^{N} C_j(0) e^{-k_j t} \tag{6.13}$$

According to Equation (6.12), the kinetics depend in general on the preparation of the system, i.e., on the initial distribution $C_j(0)$. This dependence disappears at the fast relaxation of the matrix when the kinetics is of a single exponential character.

In experimental situations there may be different ratios between the rates of matrix relaxation and chemical transformation itself. The measurement of only $C(t)$ dependencies does not allow a simple conclusion about the processes determining the observed kinetics. To clarify this question it is necessary to conduct special additional experiments.

As an illustration of possible particular cases, let us consider a model system where the microdomain of the matrix containing reagents has only two states. The rate of the chemical reaction in each of them is equal to k_1 and k_2. The probability for transition per unit time between them is equal to a and b, respectively. We do not write down here the general solution of the task but instead restrict ourselves to discussing the most important individual cases.

1. $k_1 >> a$, $k_2 >> b$, i.e., the relaxation proceeds essentially slower than the chemical transformation. In this case, the time dependence of the concentration has the form

$$C(t) \cong C_1(0)e^{-k_1 t} + C_2(0)e^{-k_2 t}, \tag{6.14}$$

reflecting the circumstance that within the time of reagents transformation the matrix does not relax (is frozen) and the kinetics depend on the initial conditions. Equation (6.14) results from Equation (6.13) at $N = 2$.

2. a, $b >> k_1$, k_2, i.e., the rate of relaxation is high in comparison to the reaction rate. In this case, the matrix instantly relaxes to the stationary distribution. The chemical transformations occur in the equilibrium matrix and the kinetics is of a single exponential character and does not depend on the initial conditions:

$$C(t) \cong C(0)\exp\left[-\left(k_i \frac{b}{a+b} + k_2 \frac{a}{a+b}\right)t\right]. \tag{6.15}$$

On obtaining, Equation (6.15) it is taken into account that both exponents corresponding to chemical reactions in both states of the system have the same indices.

3. $k_1 >> a$, $k_2 << b$, i.e., in the first state the chemical reaction rate essentially exceeds the rate of relaxation, and in the second state the situation is opposite. Then

$$C(t) \cong \left[C_1(0) - C_2(0)\frac{b}{k_1 - b}\right]e^{-k_1 t} + C_2(0)\frac{k_1}{k_1 - b}e^{-bt}. \tag{6.16}$$

Here, the reaction and relaxation equally manifest themselves in the kinetics. At $k_1 << b$ the system passes quickly into the first state and reacts with the rate constant k_1. If $k_1 >> b$, i.e., the reaction in the first state happens faster than the transition from the second state into the first one, then just these latter transitions limit the kinetics. This is a complete analog of the situation when the kinetics is determined by the matrix relaxation.

Thus, in all cases considered, the dependencies of Equations (6.14) to (6.16) represent the sum of two exponents (two states) and — what is essential — without any qualitative peculiarities which would indicate the mechanism of the phenomena. Both the spread of the rate constant and the matrix relaxation lead to the same time dependencies of the reagents concentration on time. In a general case, the time dependence of reagents concentration (Equation (6.12)) is the sum N of damping exponents (N-states) with characteristic times τ_i. These parameters, τ_i, are the mixture of both partial rate constants k_i corresponding to different states of the matrix and the probability $w_{jj'}$ defining its relaxation.

Nevertheless, in many cases a conclusion about the mechanism of solid-phase, low-temperature reactions can be made. First, these are reactions with the H-atom transfer in crystals (see Sections 4.3 and 5.5) when the kinetics are exponential. Second, there is a good manifestation of the criteria of the tunneling transition's predominance over the classical over-barrier ones (see next section).

6.2.3 The Influence of Energy Stored in the Structural-Nonequilibrium Fragments of the Medium on the Kinetics of Exothermic Chemical Transformations

Usually chemical reactions in the solid phase are considered when assuming that the matrix is inactive. But the solid phase matrix, in which exothermic transformations occur, is not always a classical thermostat. Earlier, we considered different variants when the medium actively influences the chemical processes taking place in it. This is due to the influence of intermolecular oscillations,[19,20] the role of matrix relaxation,[15,25,31-34] and the thermal effect of the reaction.[31,32,35] It was found that the intermolecular oscillations are extremely important for tunneling reactions,[20] and as for the activation transitions it is necessary to take into consideration the energy stored in the substance.[36,37] Also, it is necessary to take into account both types of the energy stored: the thermal effect of the reaction and the energy which is released under the relaxation of the amorphous phase. Before we pass to concrete chemical reactions with the transfer of heavy atomic particles (chlorine and bromine), we clear up a question about the mechanism (tunneling or above-barrier) of such reactions.

As a rule, the presence of the low-temperature plateau in the temperature dependence of the chemical reaction rate constant was considered a guarantee of tunneling.[38] For such reactions, as the chain's growth during ethylene hydrobromination by γ-induced decomposes the HBr molecules,[39] the chain photochlorination of glassily methylcyclohexane matrices, and the chain processes of chlorination and bromination of ethylene (see Reference 40), the beginning of the low-temperature plateau is in the region $T_c \approx 40–50$ K. The mechanism of these reactions was assumed to be of a tunneling character.[38]

We estimate the temperature below which, in reactions with the transfer of atoms, the tunneling transitions prevail over the Arrhenius ones.[37,41] In the simplest case, it is necessary to compare two exponents — the Arrhenius exponent describing the above-barrier transition ($Q/k_B T$; Q is the energy of activation) and the tunneling one $\left(\frac{\pi d}{\hbar}\left(\frac{mU}{2}\right)^{1/2}\right)$; U, d are the height and width of the potential barrier overcome by the particle. Here it is taken into account that the height of the barrier in the case of tunneling noticeably differs from the activation energy. Indeed the reaction path in the case of tunneling mechanism is essentially shorter as compared to the transfer through the saddle-point; the particle moves in a short way, "cutting the angle", through the higher barrier. For the limiting temperature (T_t) we obtain the expression:

$$T_t = \frac{\hbar Q}{\pi d k_B}\left(\frac{2}{mU}\right)^{1/2} \tag{6.17}$$

It should be noted that the expression for T_t was first described in Reference 42; however, the work assumed that even on tunneling the atom moves along the reaction coordinate. The ratio $U = Q$ would correspond to this condition.

Let us obtain the limiting temperature value for reactions with the intermolecular transfer of hydrogen atoms and heavy particles. In the case of reactions with H-atom transfer, the parameter values are as follows: Q = 0.5 eV, U = 1.5 eV, d = 1 Å.[23,24] The temperature at which the tunneling begins to predominate over the activation transfer is T_t = 130 K. We notice that the value T_t is appreciably higher than T, i.e., a rather wide interval of temperatures is available where the rate constant, on the one hand, relates precisely to tunneling and, on the other, very sharply depends on the temperature.[20]

Now we estimate the limiting temperature for the ethylene chlorination reaction.[40] In this review, the required parameters are given: the activation energy (Q = 1 Kcal/mol), the height (U = 1.6 Kcal/mol), and the width d = 0.5 Å of the potential barrier as well as the effective mass of the particle performing the transition ($m = 17.3\, m_H$; m_H is the H-atom mass). The corresponding limiting temperature is equal to T_t = 24 K,[41] i.e., the tunneling transfer of the chloride atom begins to predominate only at temperatures noticeably lower than $T_c \approx$ 40 to 50 K. We note that this temperature corresponds to the begining of the low-temperature plateau. Hence, the low-temperature plateau beginning at essentially higher temperatures cannot be explained by tunneling. In other words, it is necessary to look for a new explanation of the regularities of the temperature dependence of the solid-phase rate constant of chemical reactions with a heavy atoms transfer.

The interpretation of experimental data obtained in studying the low temperature reactions with the participation of heavy atoms is based on the following model.[36,37] Account is taken of the important role of the energy stored in the amorphous substance as well as the energy released as a result of the reaction. The energy of the chemical transformation initiates readjustment of structurally nonequilibrium fragments of the matrix (since in the glass substances the matrix is not in the condition with a minimum energy value) at which the stored "strain energy" is released. This energy initiates exothermic reactions which contribute to the heating of the environment and can cause a further release of the energy.

The task was analyzed within the Kramer's classical model about the averaged time of reaching the boundary in the energy domain.[36,37] A release of the particle from the potential well was considered (Figure 6.7). Apart from the equilibrium activation transitions, account is taken of the possibility for relatively rare direct transitions caused by thermal fluctuations which can call for dissociation of the coupling studied. Both exothermic reactions and the stress energy are the cause of such thermal fluctuations. For sufficiently deep potential wells, it is possible to get a simple analytical expression for the chemical process rate constant:[36,37]

$$k(T) = \frac{\kappa\Phi\left(\kappa\eta^{-1}, 1; Q/k_BT\right)}{\Phi\left(\kappa\eta^{-1}, 1; Q/k_BT\right) - 1} \approx \kappa + \eta\frac{Q}{k_BT}\exp\left[-Q/k_BT\right]. \qquad (6.18)$$

Here, $\Phi(a,b;z)$ is the degenerate hyper-geometric function; κ is the rate constant of the direct transition due to the thermal effect of the reaction and the energy stored in the medium; the coupling of the considered reaction center with the environment is described by means of the "friction" coefficient η; Q is the activation energy equal to the depth of the potential well (see Figure 6.7). In Equation (6.18), the finite low-temperature limit of the chemical reaction rate is described by means of the first term. Under the increase in temperature, the dependence becomes the activation character, the same as in the Kramer's classical model.

In Reference 37, Equation (6.18) is compared with the experimental data (see Figures 6.8 to 6.10).[39,40] The fitting parameters were selected in such a way that the deviation of the theoretical curve from the experimental dots would be minimal. It turned out that at reasonable parameters describing the environment (η) and the chemical reaction (Q) a good agreement of theoretical and experimental data is available.[37] At the same time, it is impossible to describe by Equation (6.18) the experimental data on reactions with hydrogen atoms transfer[23,24,43-45] and polymerization of formaldehyde.[46] This could be expected since the reactions mentioned have another mechanism.

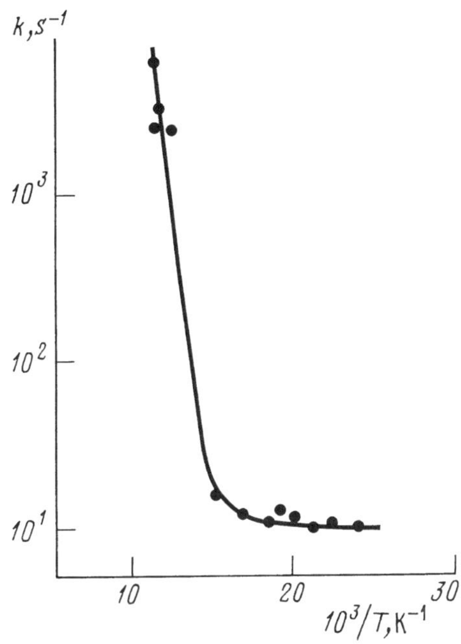

Figure 6.8 Temperature dependence of the rate constant of the ethylene hydrobromination.[39] The values of the fitting parameters: Q = 3.5 kcal/mol, $k/\eta = 10^{-10}$. • = experiment; — = fitting curve.

Apart from reactions,[39,40] the analysis of studies on photochlorination of butyl chlorine in glasses[47] is of interest, too. For this reaction, in a rather wide range of temperatures and ratios of concentrations between the reagents, the measurements of different physical and chemical magnitudes characterizing the process including the temperature dependence of the rate constant (Figure 6.11) are carried out.

The experiments show[47] that the transition to the low-temperature plateau occurs in the region T_c = 80 K. This temperature is noticeably higher than the similar value in crystals. But according to the tunneling mechanism reported in Equation 40, because of the absence of structural provision in the amorphous phase, the tunneling must appear at essentially lower temperatures than in crystals. At the same time, the above-mentioned[36,37] mechanism, due to the contribution to the reaction of the energy stored up in the amorphous phase, yields the temperature T_c close to the observed one. Moreover, the concentration dependence of the kinetics (Figure 6.11) cannot

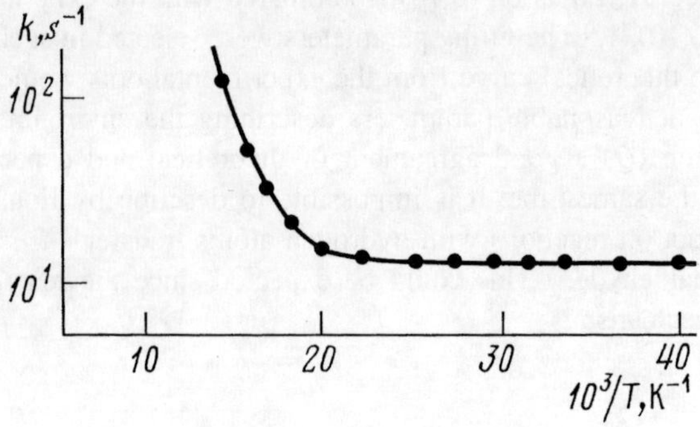

Figure 6.9 Temperature dependence of the rate constant of the ethylene chlorination.[40] The values of the fitting parameters: Q = 1.4 kcal/mol, k/η = $5 \cdot 10^{-5}$. • = experiment; — = fitting curve.

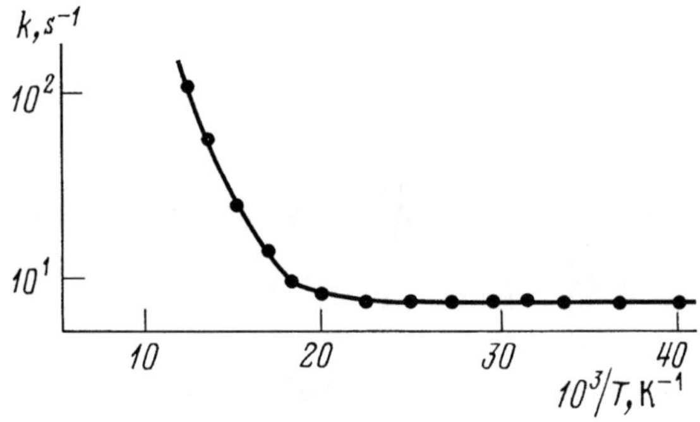

Figure 6.10 Temperature dependence of the rate constant of the ethylene bromination.[40] The values of the fitting parameters: Q = 1.4 kcal/mol, $k/\eta = 10^{-4}$. • = experiment; — = fitting curve.

be explained on the basis of the tunneling mechanism since at small concentrations (<0.1) the low-temperature plateau is absent. The concentration of reagents is very important only when the thermal effect of the process plays the main role. Therefore, the obtained concentration dependence also proves the reasonableness of the above-barrier mechanism discussed.[36,37]

The treatment of the experimental data by Equation (6.18) yields the following values of the fitting parameters: Q = 0.28 eV, $\kappa/\eta = 2 \cdot 10^{-15}$.[47] The theoretical curve (Figure 6.11, curve 1) satisfactorily describes the experiment at high and low temperatures, and in the transition region the difference is obvious. This is explained by use, in obtaining Equation (6.18), of the direct transition rate constant κ independent of temperature (more precisely on the value of the thermal energy stored up in the reaction center) which holds only when one mechanism prevails to reach the boundary (either thermal or direct).

In the case of small chlorine concentrations (Figure 6.11, curve 2), when no low-temperature limit is observed, the reaction rate constant value decreases and the activation energy becomes larger (Q = 0.38 eV). Indeed, at the decrease in chlorine concentration the matrix is more uniform and, hence, more rigid which in its turn

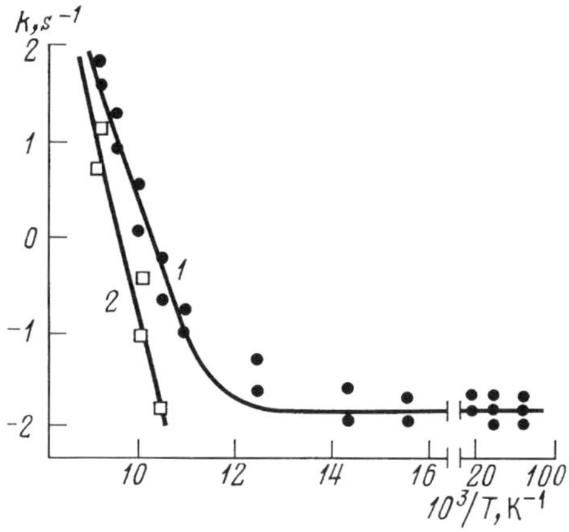

Figure 6.11 Temperature dependence of the rate constant of the butyl chlorine chlorination reaction.[47] Curve 1 corresponds to chlorine concentration, C = 0.23, and straight line 2 corresponds to a chlorine concentration of C = 0.01, Points and squares are the experiment, continuous lines are the fitting curve. The value of fitting parameters are given in the text.

leads to steristic difficulties that arise in the solid phase. The same reason stipulates the larger value of the activation energy.

The afore-mentioned consideration allows the basic regularities of low-temperature reactions with the heavy particles transfer to be described. However, it does not account for the process of energy delivery to the reaction center. For this purpose, one should write the equation of temperature conductivity taking into account the energy being released in the process of the reaction and matrix relaxation. Besides, it is necessary to utilize two more equations describing the kinetics of energy release in chemical reactions and relaxation processes.

It is this approach that was suggested in Reference 48. The realization of this model may be found in recent work in which dependencies of the reaction rate on the concentration of active particles and the reaction thermal effect were obtained.[49] These dependencies are of a barrier character and determined by the properties of the medium and molecules participating in the reaction.

The consideration undertaken in References 48 and 49 undoubtedly is a step forward as compared with the results of References 36 and 37. But experimental conditions are possible at which such an approximation is insufficient. The problem is that in describing a thermal chemical process the idea of temperature is used. At the same time,[48] the rates of relaxation processes and thermalization are comparable. In other words, the lifetime of highly excited molecules with regard to reaction and relaxation is comparable with the time required for establishing the thermal equilibrium. Therefore, to use the idea of temperature in such tasks is not entirely valid.

Thus, different mechanisms of the low-temperature limit of the chemical reaction rate exist. As we saw, only light particles — hydrogen atoms — tunnel. As far as heavy particles are concerned, in some cases the tunneling is a compound part of the complex chemical process: the tunneling rotation of reagents plays a part (see Section 4.3.1). In other cases, the reaction mechanism is of an above-barrier character and the parameters of the curves describing the reaction kinetics depend both on reagent properties and the geometrical, dynamic, and relaxation characteristics of the environment.

6.3 COMPENSATION EFFECT AND CAGE EFFECT

When radiation-chemical and other solid phase reactions take place one often is faced with a phenomena known for a long time but for which a complete understanding of their nature is not available.

In 1926, an unusual behavior of the activation energy and pre-exponential factor for the rate constant of the catalytic oxidation reaction was discovered.[50] These values do not remain constant at variations of temperature, but change simultaneously. That is, the decrease or increase in the activation energy is compensated for by the corresponding change of the pre-exponential factor. Further, similar properties appeared in homogeneous monomolecular chemical reactions, too.[51] This effect, later called compensational, was observed in different chemical processes including the reactions of a free radical recombination (see next section).[52]

The nature of this phenomena has been discussed repeatedly in scientific literature[53] from the standpoint of the classical mechanism of the reaction elementary act. The characteristic exceptions of the compensation effect are detected. At the same time, the observed regularities for reactions in low and high molecular compounds are not explained. Apparently this is related to the nonclassical way of overcoming the potential barrier by the particle. The temperature dependence of the transition rate constant of Equations (4.7) and (4.9) is qualitatively different from the Arrhenius one. In the following section it is shown by the example of the radical recombination reaction in polymers (see Table 6.1) how the experimental results are explained in a rather simple way under the reasonable values of the parameters.

Table 6.1 Parameters of the Potential Barrier and Media Received by Fitting Procedure

Polymer	Region of recombination	E_a (eV)	K_0 (cm/s)	R_0 (Å)	$\gamma \times 10^{-4}$ (din/cm)	V_0 (eV)
High-density polyethylene	Amorphous	0.03	1.4×10^{-15}	3.32	16.6	0.73
	Surface of lamellas	1.00	1.8×10^{13}	3.94	1.62	2.49
	Inner part of lamellas	1.08	1.8×10^{10}	3.67	3.16	1.83
Low density polyethylene	Amorphous	0.02	2.8×10^{-16}	3.30	30.4	0.67
	Surface of lamellas	0.41	5.6×10^{-4}	3.47	4.59	1.24
	Inner part off lamellas	0.80	16	3.44	6.68	1.11
Polypropylene	Isotactic	—	—	3.85	4.86	2.26
	Syndiotactic	—	—	3.85	4.78	2.26
	Atactic	—	—	4.01	2.61	2.63

Another interesting exception to the radiation-chemical process is the effect of the molecular environment and temperature on the possible decay reactions of excited molecules in the solid phase. More refined effects of the environmental influence on the temperature dependence of the rate constant of radicals photodissociation will be considered in the following chapter. Here, we shall only point out the principal possibility of the exclusion of excited molecules dissociation.

In fact, under the dissociation of the molecules in two radicals in order to avoid a fast recombination, one of them should leave the place of creation. The radicals in the solid phase, as a rule, are not able to disperse and recombine ("cage" effect). The hydrogen-containing molecules dissociate successfully with the decomposition of the *R-H* bond since the hydrogen atoms readily diffuse in the solids. This circumstance explains the primary role of the H-atom split-off from molecules in the solid phase and justifies the great attention devoted to this question. The dissociation of paramagnetic particles (molecular ions, radicals) takes place in the solid phase, not only with obstruction of the H atom but also of large fragments, since the number of the coupled electrons is not increased in this process.

Naturally, the role of the "cage" increases with the decrease in temperature. This is easy to understand if one considers that to transfer a large molecular fragment a channel must be formed in the medium; i.e., the molecules surrounding the dissociated reagent must diverge and let the obstruction fragment “pass”. The probability for this event or, in other words, the amplitude of intermolecular oscillations depends on temperature. At low temperatures, this probability is small which ensures the observance of the "cage" effect.

6.4 RECOMBINATION KINETICS OF FREE RADICALS IN THE SOLID PHASE

The recombination reaction is an important radiation-chemical reaction determining the lifetime of free radical in the solid phase at low temperatures. The studies of radical recombination in the solid phase have been carried out for a long time, and a number of models have been suggested (see Section 3.3); however, up to now there has been no uniform judgment of its mechanism. First of all, it is necessary to explain the mechanism of the motion of free radicals in the solid phase at low temperatures. For recombination, the encounter of radicals is necessary and it was usually assumed that it takes part as a result of molecular motions.

It is clear that at low temperatures the motions of the radicals are hindered; therefore, a mechanism of mobility was suggested, stipulated by the transfer of the H atom from a molecule to radical.[54] There is not a transfer of a radical over the substance as a whole, but a migration of free valence as a result of the reaction:

$$\mathrm{H} + \mathrm{H}R \rightarrow R\mathrm{H} + R \tag{6.19}$$

The radical and the molecule change places which corresponds to the transfer of the radical.

A transition of the hydrogen atom is possible by either the above-barrier or tunneling method. The processing of the experimental data by formulae corresponding to the above-barrier transition led to the anomalous values of the pre-exponential multiplier and to a noticeable dependence of the activation energy on temperature. To explain these regularities, an effort was made to use the notation of the compensation effect.[52,53]

In 1978, a tunneling mechanism of free valence migration was suggested under the assumption that the temperature dependence of the process rate constant is determined by intermolecular oscillations as a result of which the distance between reagents changes (see Chapters 4 and 5).

We consider the kinetics of radical recombination in the solid phase. It is known that it has a stepped character.[56] The concentration of radicals at constant temperature first decreases with time, and then their annihilation slows down and after some time comes to a stop. With the increase in temperature, the process restores (Figure 6.12). This can be explained if account is taken that for recombination the mobility of radicals is important and is stipulated by the process in Equation (6.1). In case of the tunneling mechanism of H-atom transfer, the rate constant exponentially depends on temperature (see Equation (4.7)). The parameters of this dependence are determined by the characteristics of the molecular environment of the radical. It is obvious that the regions of temperatures affecting the potential barrier parameters at which the motions defrost can be different.

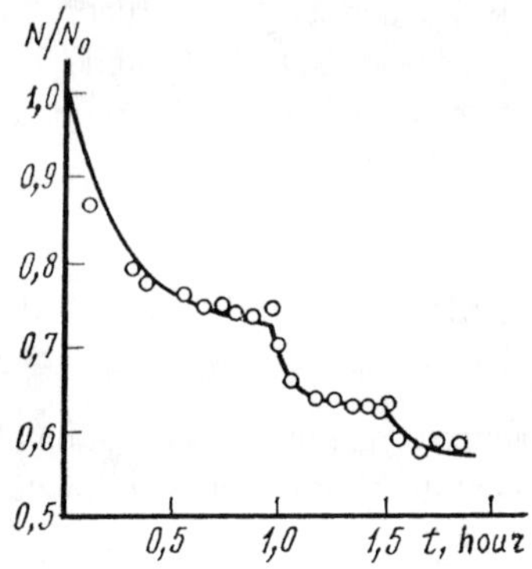

Figure 6.12 Stepped recombination of free radicals in phenol.[56]

In considering the kinetic dependencies, it is necessary also to take into consideration the character of the ionizing radiation passing through the substance. As a result of the radiation effect in the sample, a sharply nonuniform distribution of radicals in the domain is created (spurs). If it is assumed that the radicals in the spur are distributed according to Gauss, the dimensions of the spurs are determined by diffusion, and the number of radicals by recombination, then for the number of radicals in the spur (N) we obtain:[57]

$$N(t) = \frac{N_0}{1 + \alpha_0\left[1 - (1 + t/\tau)^{-1/2}\right]},$$

$$N_0 = N(t = 0), \qquad \tau = \sigma^2/2D.$$

Here, σ is the square-root initial dimension of the spur, $D = kd_0^2/6$ is the diffusion coefficient. The rate constant k is determined by Equation (4.7) (see also Equation (4.9)), d_0 is the "transfer" of the radical at each transition of the H atom, the parameter a_0 depends on the characteristics of the system.[57]

Figure 6.13 shows the experimental data on the stepped recombination.[58,59] It is easy to see that the stepped character of recombination can be explained by a sharp increase in the diffusion coefficient of free valencies under increased temperature.

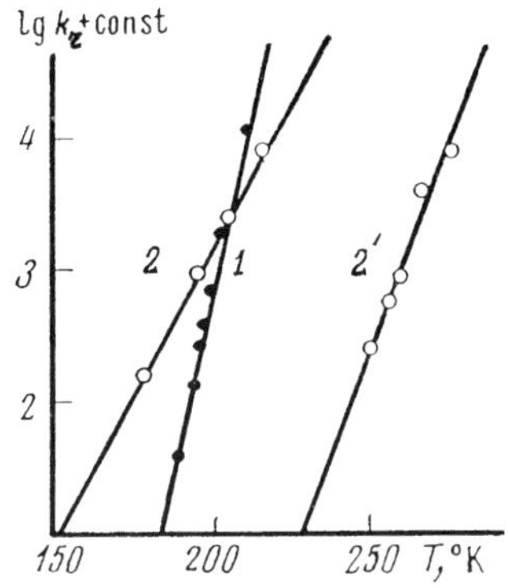

Figure 6.13 Dependence of recombination rate constant on temperature. 1, in glycerine,[58] 2,2′, in polypropylene.

The given mechanism satisfactorily describes the experimental data on the temperature dependence of the rate constant of radicals recombination.[60] For the barrier which is obtained at imposing two potential curves corresponding to the ruptured and created bonds, a comparison of the suggested model with literature experimental data was carried out.[61,62] These works investigated the kinetics of radical recombination in polyethylene and polypropylene. Several temperature regions of accelerated recombination responsible for recombination in the amorphous phase on the surface and inside lamellas were found. In each phase, the kinetics obeyed the secondary-order law.

Now let us see to what values of the independent parameters: R_0 (equilibrium spacing between reagents) and γ (force constant describing the coupling between the reagents) the suggested model leads. By comparing the obtained temperature dependencies of the rate constant in different ranges in Equation (4.7), we can find the above-mentioned parameters, which are given in Table 6.1.[61,62] The letter U_0 denotes the height of the potential barrier at the given distance R_0.

For polyethylene in the amorphous phase, the force constant turned out to be maximum and the distance between the radical and molecule–minimum. This testifies to the possibility of the fragments of molecules coming close and a strong bond, CH–C, in the amorphous phase created. In the crystal phase, the approach of fragments is hindered and as a result, a weaker bond is formed having less force constant and a larger length. On the surface of the lamellas, where the addition of molecules takes place, the length of the coupling turned out to be larger than inside the lamellas. The lengths of the couplings in polyethylene of high density are less than in polyethylene of low density. The lengths of the couplings and force constant obtained in processing the data for polypropylene are approximately equal for isotactical and syndiotactical polypropylenes. For the disordered atactical propylene, the coupling length turned out to be somewhat larger and can be explained by the packing of less density.

For comparison, Table 6.1 gives the results of processing the experimental data on the Arrhenius dependence of the rate constant on temperature:

$$k(T) = k_0 \exp[-E_a/k_B T].$$

It is seen that such a consideration leads to the anomalous values of the pre-exponential factor and unusually low energy of activation for the recombination radicals in the amorphous phase.

It is necessary to note that the obtained lengths and force constants by the order of magnitude coincide with the lengths and force constant of the usual hydrogen coupling ($R_0 = 3$ Å and $\gamma = 3 \cdot 10^4$ din/cm) which testifies to a certain similarity of the coupling of the radical with molecule and the usual hydrogen bond.

The experimental evidence of the tunneling mechanism of the migration of free valences at the recombination of radicals was obtained later.[63] The low-temperature recombination of free radicals in usual and deuterized polyethylene was investigated. To process the experimental data, Equation (4.7), describing the dependence of the tunneling transition rate constant of hydrogen atom on temperature was utilized.

Figure 6.14 shows the kinetic curves of the annihilation of free radicals in the irradiated polyethylene and deuterized polyethylene (98% of isotope replacement) at 77 K. During the experience the radicals stabilized in the amorphous phase of the usual polyethylene recombine. Their quantity makes up 12% of the total quantity of radicals. In the deuterized polyethylene and within the same time, the recombination of radicals was not observed. Thus the isotope effect demonstrates that the recombination is stipulated not by the diffusion of free radicals together with the polymer chain when the deuterization will play no role but rather by the migration of free valences at the transfer of hydrogen atom (deuterium) from the molecule to radical.

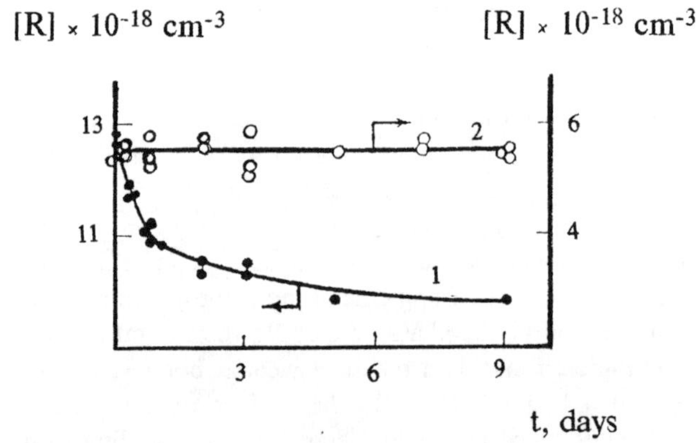

Figure 6.14 Kinetic curves of free radical recombination in irradiated polyethylene (1) and deuterated polyethylene (2) at 77 K.

However, the observed isotope effect does not make it possible to answer the question of what mechanism (above-barrier or tunneling) is utilized to realize the transition. Therefore, measurements of the dependence of the radical recombination rate constant on temperature were undertaken. The processing of the experimental data in conformance with the Arrhenius law leads to a small value of activation energy ($3.8 \cdot 10^3$ J/mol) and an anomalously small value of the pre-exponential factor ($4.5 \cdot 10^{-27}$ m^3/spin/sec). The same data were processed in accordance with Equation (4.7). The required potential surface for the tunneling hydrogen atom was calculated by the LEPS method.

In Figure 6.15 the solid curve shows the calculated dependence of the recombination rate constant on temperature under the following parameter values: the distance between the molecule and radical is $R_0 = 3.5$ Å and the corresponding force constant of the coupling is $\gamma = 3.45 \cdot 10^4$ din/cm. At these parameters, agreement between theory and experiment was observed. The obtained parameters corresponding to the recombination in the amorphous phase of polyethylene agree with similar parameters found earlier for other phases of polyethylene.[60] It follows, then, that the most probable mechanism of free radical recombination in polyethylene is the migration of free valence in tunneling a hydrogen atom from the molecule to radical.

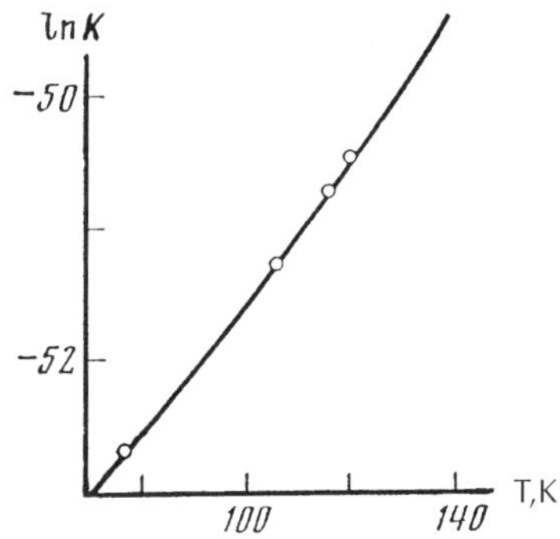

Figure 6.15 Calculated dependence of free radical recombination rate constant in irradiated polyethylene on temperature (solid curve) and experimental points.

The rate of radical recombination essentially increases at temperatures close to the temperature of the phase transition. This agrees with the considered tunneling mechanism of recombination. As was shown in Reference 63, in the vicinity of the phase transition the amplitudes of molecule oscillation abruptly increase which, as mentioned above, leads to the increase in the probability for the tunneling transition.

Recently, direct evidence of the tunneling migration of the free valence in the solid matrix has been obtained. The recombination of hydrogen atoms at the temperature of liquid helium in γ-irradiated molecular hydrogen was investigated. The initial concentration of hydrogen atoms was $1.3 \cdot 10^{-7}$ mol/cm^3. A decrease in the concentration of hydrogen atoms of two times took place up to 30 hours. The kinetics of their annihilation obeyed the second-order law and was controlled by diffusion. Two possible mechanisms for the free valences diffusion were discussed: first, the immediate migration of the hydrogen atom (from trap to trap) and, second, the transfer of the free valence as a result of the reaction $H + H_2 \rightarrow H_2 + H$ by the tunneling mechanism. It has been shown that the second tunneling mechanism of the migration of free valence in the solid phase at low temperatures is realized.

REFERENCES

1. Jackson, J.L., Montroll, E.W., Free radical statistics, *J. Chem. Phys.,* 28, 1101, 1958.
2. Golden, S., Free-radical stabilization in condensed phases, *J. Chem. Phys.,* 29, 61, 1958.
3. Vasil'ev, G.K., Tal'rose, V.L., On the theory of stabilized radicals accumulation in solids, *Kinet. Catalysis,* 4, 497, 1963.
4. Alger, R.S., Anderson, T.H., Webb, L.A., Irradiation effects in simple organic solids, *J. Chem. Phys.,* 30, 695, 1959.
5. Siegel, S., Baum, L.H., Skolnik, S., Flournoy, J.M., Observations of the thermal behavior of radicals in gamma-irradiated ice, *J. Chem. Phys.,* 32, 1249, 1960.
6. Plotnikov, V.G., Ovchinnikov, A.A., Photo- and radiation-chemical molecular stability. Monomolecular reaction of hydrogen atom abstraction, *Russ. Chem. Rev.,* 47, 444, 1978.
7. Trakhtenberg, L.I., Milikh, G.M., Kinetics of free radicals accumulation on irradiation of solids, *High Energy Chem.,* 17, 483, 1983.
8. Kabi, A., Free-radical in cyclohexane irradiated at 77 K, *Int. J. Rad. Phys. Chem.,* 11, 49, 1969.
9. Sauer, M.C., Mani, I., Pulsed radiolisis of liquid cyclohexane and n-hexane. I. Yield of hydrogen atom measured by the cyclohexadienil radicals absorption in solution containing bezene. II. Absorption spectra and reactions of cyclohexyl and hexyl radicals, *J. Phys. Chem.,* 72, 3856, 1968.
10. Yakimchenko, O.E., Doroshina, G.P., Lebedev, Ya.S., ESR spectra and structure of radical pairs in γ-irradiated dimethylglyoxime single crystals, *High Energy Chem.,* 3, 242, 1969.
11. Plonka, A., *Lecture Notes in Chemistry,* Springer-Verlag, Berlin, 1986.
12. Vyazovkin, V.L., Bolshakov, B.V., Tolkatchev, V.A., H-atom abstraction from alcohol by alkyl radicals. Studies of the limiting stage, *Chem. Phys.,* 95, 93, 1985.
13. Plonka, A., Developments in dispersive kinetics, *Progr. React. Kinet.,* 16, 157, 1991.
14. Vyazovkin, V.L., Bolshakov, B.V., Tolkatchev, V.A., Kinetics of methyl radical reaction with methanol at 20-105 K, *Chem. Phys.,* 75, 11, 1983.
15. Tolkatchev, V.A., The effect of local environment relaxation on elementary homolitic reaction kinetics in glass. Interpretation of the Kohlrausch-Williams-Watts relaxation function, *Chem. Phys.,* 116, 283, 1987.
16. Bolshakov, B.V., Stepanov, A.A., Tolkatchev, V.A., The $\sqrt{t}$ law in solid-phase reactions of the $\sqrt{t}$ law stability in H-atom abstraction reactions, *Intl. J. Chem. Kinet.,* 12, 271, 1980.
17. Zaskulnikov, V.M., Vyazovkin, V.L., Bolshakov, B.V., Tolkatchev, V.A., The $\sqrt{t}$ law in solid-phase reactions. Analysis of hypotetic on rate constant distributions, *Intl. J. Chem. Kinet.,* 13, 707, 1981.
18. Burshtein, A.I., Klochikhin, V.L., Trakhtenberg, L.I., Polychromatic kinetics of cagial reactions in solid phase, *Sov. J. Chem. Phys.,* 3, 155, 1984.
19. Klochikhin, V.L., Trakhtenberg, L.I., Role of the active particle migration in the cagial radical reactions, *Sov. J. Phys. Chem.,* 58, 2877, 1984.
20. Goldanskii, V.I., Trakhtenberg, L.I., Fleurov, V.N., *Tunneling Phenomena in Chemical Basics,* Gordon and Breach Science, New York, 1989.
21. Doba, T., Ingold, K.U., Sibrand, W., Evaluation of reaction rate constants in media where they show dispersion: hydrogen abstraction by methyl radicals in a methanol glass, *Chem. Phys. Lett.,* 103, 339, 1984.
22. Willard, J.E., Some properties of trapped electrons, radicals and H-atoms in glassy hydrocarbons, *Intl. J. Rad. Phys. Chem.,* 6, 325, 1974.

23. Hudson, R.L., Shiotany, M., Williams, F., Hydrogen atom abstraction by methyl radicals in methanol glasses at 15-100 K. Evidence for a limiting rate constant below 40 K by quantum mechanical tunneling, *Chem. Phys. Lett.,* 48, 193, 1977.
24. LeRoy, R.J., Murai, H., Williams, F., Tunneling model for hydrogen abstraction reactions in low-temperatures solids. Application to reactions in alcohol glasses and acetontril crystals, *J. Am. Chem. Soc.,* 102, 2325, 1980.
25. Tolkatchev, V.A., Syutkin, V.M., Vyazovkin, V.L., The role of local motions in the kintic peculiarities of atomic tunneling in solids, *Chem. Phys.,* 170, 427, 1993.
26. Klochikhin, V.L., Trakhtenberg, L.I., Polychronic kinetics of cage reactions in solid phase, *Sov. J. Phys. Chem.,* 62, 2345, 1988.
27. Beklemishev, S.A., Slavinskaya, N.A., Trakhtenberg, L.I., Polychronic kinetics of ethyl radicals reactions with tryethylamine, *High Energy Chem.,* 22, 89, 1988.
28. Trakhtenberg, L.I., Klochikhin, V.L., Pshezhetskii, S.Ya., Tunneling of hydrogen atom in low temperature processes, *Chem. Phys.,* 59, 191, 1981.
29. Goldanskii, V.I., Kozhushner, M.A., Kinetics of polychromatic chemical reactions with mixing according to the rate constants, *Dokl. Phys. Chem.,* 274, 632, 1984.
30. Berezhkovskii, A.M., Trakhtenberg, L.I., Polychronic kinetics of chemical reactions relaxing matrix, *Dokl. Phys. Chem.,* 303, 118, 1988.
31. Ovchinnikov, A.A., Onischuk, V.A., Kinetics of chemical reactions in glasses, *Sov. J. Chem. Phys.,* 3, 511, 1984.
32. Berezhkovskii, A.M., Trakhtenberg, L.I., Chemical reactions in solid matrix induced by external action, *Dokl. Phys. Chem.,* 275, 1438, 1984.
33. Rendell, R.W., Interaction of relaxing system with a dynamical environment, *Phys. Rev. E.,* 48, R17, 1993.
34. Wang, J., Wolynes, P., Survival path for reaction dynamics in fluctuating environments, *Chem. Phys.,* 180, 1141, 1994.
35. Semenov, N.N., Collective interactions in polymerisation processes at low temperature and in polymers with conjugated bonds, *J. Polym. Sci.,* 55, 563, 1961.
36. Timashev, S.F., Intermittency in chemical reactions kinetics in solids, *Russ. J. Phys. Chem.,* 66, 846, 1992.
37. Timashev, S.F., Trakhtenberg, L.I., Classical above-barrier mechanism of low temperature limit of chemical reaction rates, *Russ. J. Phys. Chem.,* 67, 401, 1993.
38. Benderskii, V.A., Makarov, D.E., Wight, C.A., *Chemical Dynamics at Low Temperatures,* John Wiley & Sons, New York, 1994.
39. Barkalov, I.M., Goldanskii, V.I., Kiryukhin, D.P., Zanin, A.M., Kinetics and mechanism of the low temperature chain hydrobromination of ethylene, *Chem. Phys. Lett.,* 73, 273, 1980.
40. Benderskii, V.A., Goldanskii, V.I., Tunneling of heavy particles in the low temperature chemistry, *Intl. Rev. Phys. Chem.,* 11, 1, 1992.
41. Trakhtenberg, L.I., Mechanisms of low temperature transition of atoms, *Russ. J. Phys. Chem.,* 69, 1995.
42. Goldanskii, V.I., Role of the tunneling effect in the chemical reactions kinetics at low temperatures, *Dokl. Phys. Chem.,* 124, 1261, 1959.
43. Toriyama, K., Nunome, K., Iwasaki, M., Evidence for tunneling hydrogen atom transfer reactions at 4.2 K in organ crystals, *J. Amer. Chem. Soc.,* 99, 5823, 1977.
44. Prass, B., Colpa, J.P., Stehlik, D., Identification of the lowest energy nuclear fluctuation modes promoting the photochemical H-transfer tunneling reaction in doped fluoren single crystals, *J. Chem. Phys.,* 88, 191, 1988.
45. Prass, B., Colpa, J.P., Stehlik, D., Intermolecular H-tunneling in solid state photoreactions promoted by distinct low-energy nuclear fluctuation modes, *Chem. Phys.,* 136, 187, 1989.

46. Goldanskii, V.I., Frank-Kamenskii, M.D., Barkalov, I.M., Quantum low temperature limit of a chemical reaction rate, *Science,* 182, 1344, 1973.
47. Benderskii, V.A., Misochko, E.Ya., Filippov, P.G., Kinetics pecularities of the chain chemical reactions in the glass region of the reagents mixture, *Sov. J. Chem. Phys.,* 5, 955, 1986.
48. Timashev, S.F., Tovbin, Yu.K., Trakhtenberg, L.I., The influence of nonlinearity of a medium on kinetics of solid-state processes, *Russ. J. Phys. Chem.,* 68, 1949, 1994.
49. Trakhtenberg, L.I., Tovbin, Yu.K., Timashev, S.F., Votyakov, E.V., The problems of low temperature chemical kinetics in relaxing solid phase matrix, in *Modern Trends in Low Temperature Chemistry,* Moskow University Publishing House, 1994, 32.
50. Constable, F.H., The mechanism of catalytic decomposition, *Proc. Roy. Soc.,* A108, 355, 1925.
51. Roginsky, S.Z., Rozenkevitsch, L.V., Zum aufbay der guanten-theorie der chemischen kinetik, *Z. Phys. Chem.,* B10, 47, 1930.
52. Burshtein, A.I., Molecular-kinetics aspects of condensed state chemical physics, *Russ. Chem. Rev.,* 47, 212, 1978.
53. Frank, J., Rabinowitch, E., Some remarks about free radicals and the photochemistry of solutions, *Trans. Faraday Soc.,* 30, 120, 1934.
54. Dole, M., Keeling, C.D., Long range migration of chemical activity in the solid state, *J. Amer. Chem. Soc.,* 75, 6082, 1953.
55. Klochikhin, V.L., Pshezhetskii, S.Ya., Trakhtenberg, L.I., On the theory of free radicals recombination in solid state, *Dokl. Phys. Chem.,* 239, 324, 1978.
56. Mikhailov, A.I., Lebedev, Ya.A., Buben, N.Ya., "Stepped" recombination of free radicals in irradiated organic substance, *Kinet. Catalysis,* 5, 1020, 1964.
57. Samuel, A.H., Magee, J.L., *J. Phys. Chem.,* 21, 1080, 1953.
58. Ermolaev, V.K., Molin, Yu.N., Buben, N.Ya., Radical recombination in solid organic substance, *Kinet. Catalysis,* 3, 314, 1962.
59. Nara, S., Kashiwabara, N., Sohma, J., Configurations of the free radicals in irradiated polypropylene and the relation of their decay reactions to the molecular motion, *J. Polym. Sci.,* A5, 929, 1967.
60. Klochikhin, V.L., Pshezhetskii, S.Ya., Trakhtenberg, L.I., Tunneling mechanism of free valences mobility in polymers, *Polym. Sci.,* A21, 2792, 1979.
61. Nara, S., Shimada, S., Kashiwabara, N., Sohma, J., Relation of the decay of free radicals in irradiated polyethylene to the molecular motion of the polymer and the configuration of the free radicals, *J. Polym. Sci.,* A6, 1435, 1968.
62. Kusumoto, N., ESP studies of polypropylene with different stereospecificities, *J. Polym. Sci.,* 837, 1968.
63. Klochikhin, V.L., Pshezhetskii, S.Ya., Slavinskaya, N.A., Trakhtenberg, L.I., About mechanism of low temperature recombination of free radicals in polyethylene, *Polym. Sci.,* B23, 855, 1981.
64. Miyazaki, T., Lee, K.-P., Fueki, K., Takcuchi, A., Temperature effect on the decay of H(D) atoms in the radiolisis of solid H_2, D_2 and HD at 4.2 K and 1.9 K. Evidence for tunneling migration, *J. Phys. Chem.,* 88, 4959, 1984.
65. Miyazaki, T., Iwata, N., Lee, K.-P., Fueki, K., Decay of H(D) atoms in solid hydrogen at 4.2 K. Rate constant for tunneling reaction $H_2(D_2,HD)+H(D)$, *J. Phys. Chem.,* 93, 3352, 1989.

7 PHOTORADIATION PROCESSES

In this chapter we consider some effects related to the mutual action of high energy radiation and light on matter. The processes accompanying these effects were given the name "photoradiational".[1,2] It was found that in a condition of mutual irradiation, the effects of non-additivity are present. In other words, the changes taking place under the simultaneous effect of radiation and light cannot be reduced to summation of the actions of these factors apart. In considering the processes taking place under the mutual action of the high energy radiation and light, it is not necessary to consider the regularities of the elementary act. In this case, it is necessary to determine what known elementary acts lead to the experimentally observed nonadditivity.

The basic techniques used in the investigation of photoradiation processes are ESR chromatography and optical spectroscopy. The investigations are conducted both at the stage of creation of reaction products and at the stage of transformation of intermediate short-lived particles. Investigation of the intermediate stage is preferable, since it allows the mechanism of the photoradiation process to be explained. A criterion of the effectiveness of these processes can be the disintegration of molecules into fragments: atoms, ions, radicals. The paramagnetic properties of these particles allow the processes of their creation and annihilation at such interactions to be followed. Using the ESR spectra it is possible to determine the type and concentration of the intermediate particles stabilized at low temperatures.

Under mutual irradiation, two main events can be realized: (1) the light is absorbed by the molecules of the base matter, and (2) the light is absorbed only by short-lived intermediate particles arising under the effect of ionizing radiation (radicals, stabilized electrons, and molecules excited by electron impact). The absorption band of these intermediate products of radiolysis is shifted to the long-wave region. Naturally, it turns out to be preferable to study a simpler event, such as (2), when it is relatively easy to separate the interaction of the visible radiation with short-lived particles.

In these investigations the following types of irradiation were used:

1. Irradiation solely by ionizing radiation
2. Consecutive irradiation — that is, irradiation at which a sample is exposed to ionizing radiation and then for the same period of time exposed to light.

3. Simultaneous irradiation — that is, a period of irradiation during which the matter is under the effect of ionizing radiation and light for the entire exposure time.
4. Cyclic irradiation — that is, irradiation during which the matter is exposed to repeated cycles of consecutive irradiation.

7.1 DEPENDENCE OF THE RADICAL CONCENTRATION ON THE METHOD OF IRRADIATION

Reference 2 cites experimental data on the processes in matter under mutual exposure to ionizing radiation and light. In particular, the data are given showing that at the different regimes of irradiation the efficiency of radical creation is different. As a rule, the inequality is realized:

$$R_e > R_{e+h\nu} > R_{e,h\nu} \, , \tag{7.1}$$

and in some cases:

$$R_{e+h\nu} > R_e, R_{e,h\nu} \, . \tag{7.2}$$

Here, R_e is the concentration of the stabilized radicals created under the effect of ionizing radiation; $R_{e+h\nu}$, $R_{e,h\nu}$ are the concentration of radicals under simultaneous and consecutive irradiation.

The consideration carried out in Reference 3 allowed the inequality of Equation (7.1) to be explained, but it does not explain the inequality of Equation (7.2). The reason is that only the annihilation of radicals under the effect of light was taken into consideration (such a situation is realized for the majority of the investigated substances), and their creation was not considered.

It is of interest to consider a more general case when account is taken of radical creation under the absorption of light by the intermediate particles formed by the effect of radiation. This process can take place continuously during the entire time of simultaneous irradiation.

As is known, under the irradiation of alcohols a large yield of stabilized electrons is observed. Their concentration attains a stationary value. It was shown that under the effect of light the electrons disappear and the radicals appear in an equivalent quantity.[4] In this case, the effect of the simultaneous effect of ionizing radiation and light is that after the electrons captured in the traps are excited by light and react with molecules, their places can be occupied by new electrons, etc.

In addition, it is of interest to study the creation of radicals under the effect of light on short-lived intermediate particles which arise under radiation. In this case, the effect of mutual actions of radiation and light can develop, also.

The lifetime of the intermediate particles must be, on the one hand, less than the time between exposure to ionizing radiation and light under consecutive exposure and, on the other, rather large so that a sufficient concentration of this intermediate product be created.

Such a condition is met with the molecules excited to the lower triplet level; the lifetime of the molecules excited in the triplet state varies from tenths of a second to several seconds (see Section 1.3.2).[5]

In the following sections, two mechanisms of radicals creation with use of the intermediate particles will be considered in detail. The first mechanism is dissociation or any other reaction of molecules additionally excited with light to the reactivity level from the lower metastable state. The second one is the dissociation capture of the trap-excited electron by the molecule. The experimental data verifying these mechanisms are given, also.

7.2 CREATION OF THE RADICALS UNDER THE ABSORPTION OF LIGHT BY MOLECULES EXCITED WITH RADIATION TO THE LOWER TRIPLET STATE

Let us consider the mechanism of the formation of radicals caused by the effect of light on short-lived particles generated under irradiation of matter by ionizing radiation. It is known from photochemistry[7-9] that a dissolved aromatic compound under exposure to light stimulates a chemical reaction of the solvent molecules. The role of the aromatic compound molecules is reduced to the stepped absorption of the two quanta of light and the transmission of the excitation energy to the solvent molecules. The property of the aromatic compounds to have a long-lived triplet level which lies below the singlet one is utilized. The lifetime of the molecules in the lower triplet state makes up several seconds. Hence, under exposure to light a great deal of molecules excited to the lower triplet level are accumulated. Then, on absorbing one more quantum of light the aromatic compound molecules transit to the higher triplet state (T–T absorption) whose energy is enough for the ionization or dissociation of the solvent molecules.

It is quite clear that the possibility for an excited molecule to participate in the chemical reaction does not completely depend on how the aromatic compound molecule finds itself in the lower triplet state (absorption of light, electron impact, etc.). Reviews present data on the excitation of the aromatic compounds molecules to the lower triplet level under the effect of ionizing radiation.[5,6,10] The studies showed that the lower triplet level is excited effectively under the effect of radiation, with the radiation yield of the triplet states reaching 6.

Thus, under the simultaneous exposure of the solution containing an aromatic addition, a stepped excitation of the higher triplet state to ionizing radiation and light takes place. We shall note that the lower triplet level is populated under the effect of ionizing radiation, and the higher one under the following absorption of a light quantum. Let us find out how much the concentration of radicals can be expected to increase under the simultaneous irradiation with regard to irradiation by ionizing radiation only. The equations describing the kinetics of radical accumulation under the simultaneous irradiation follows:[11]

$$\frac{dn_1}{dt} = k_1 I + \left(A_{21} + B_{21} j\right) n_2 - \left(A_{10} + A_{10}^e + B_{21} j\right) n_1$$

$$\frac{dn_2}{dt} = k_2 I + B_{21} j n_1 - \left(A_{21} + B_{21} j + k_3\right) n_2 \tag{7.3}$$

$$\frac{dR_{e+hv}}{dt} = kI + k_3 n_2 .$$

Here, n_1 and n_2 are the populations of the lower and higher triplet levels of molecules; k_1 and k_2 are the rate constants of the population of these levels under the effect of ionizing radiation; A_{10} and A_{21} are the probabilities (per unit time) for the transitions $1 \rightarrow 0$ and $2 \rightarrow 1$; B_{21} is the extinction coefficient of the lower triplet level (T–T absorption); A_{10}^e is the decontamination rate of the lower triplet level with thermal electrons; k_3 is the chemical reaction rate constant with the participation of molecules excited to the higher triplet state; k is the rate constant of the radical accumulation under the effect of electrons on the solvent molecules; I, j are the intensities of ionizing radiation and light.

After time $t \sim 1/(A_{10} + A_{10}^e)$, n_1 and n_2 attain stationary values and the time of establishing the stationary concentration of the radicals essentially exceeds the lifetime in the lower triplet state. With the account that $B_{21}\, j << A_{21}$ we obtain for the concentration of the radicals at the linear part of the accumulation curve

$$R_{e+hv} \Big/ R_e = 1 + \frac{\Phi}{\left(A_{10} + A_{10}^e\right) \big/ B_{21} j + \Phi G_R}, \qquad \Phi = \frac{k_3}{A_{21} + B_{21} j + k_3}. \tag{7.4}$$

Here, Φ is the quantum yield of the chemical reaction of radical creation at the excitation of the triplet states. From Equation (7.4) it follows that the ratio R_{e+hv}/R_e depends on the relationship of the intensities I and j ($A_{10}^e \leq \sigma_{10} n_e v_e$); σ_{10} is the cross-section of the decontamination of the excited molecule under the inelastic collision with the electron; $n_e(I)$, v_e are the concentration and the velocity of thermal electrons devastating the lower triplet level. In Equation (7.4), where the absorbed dose normalization $k_1 + k_2 = G_T$ and $k_1 + k_2\, \Phi = G_R$, G_T and G_R are the yield of the triplet states and radicals under the irradiation of the solution with solely ionizing radiation.

A maximum increase in the radical concentration at simultaneous irradiation is achieved if $(A_{10} + A_{10}^e)/B_{21} j << \Phi$, then

$$R_{e+hv}/R_{e+hv} = 1 + G_I/G_R, \tag{7.5}$$

The given consideration relates not only to solutions of organic compounds; the obtained results also can be extended to any systems in which the long-lived excited states exist.

To prove experimentally the aforementioned photoradiation effect, the exceptions of radical creation under simultaneous irradiation of the aromatic compounds (benzene, naphthalene) in methylcyclohexane[11] were studied. These solutions at 77 K create a transparent matrix. The yield of the triplet states under the irradiation of benzene and naphthalene in methylcyclohexane with fast electrons is 1.5.[6]

The experimental results[11] at the simultaneous exposure of solutions to fast electrons and light and irradiation solely with fast electrons are presented in Figure 7.1. The radical accumulation curve under the simultaneous exposure of the solution to fast electrons and light lies above the radical accumulation curve under irradiation with fast electrons. For a pure solution in the absence of the aromatic admixture under simultaneous exposure to a high energy irradiation and light, the increase in the radical concentration is not observed. In both cases, under exposure of the solutions — benzene-methylcyclohexane and naphthalene-methylcyclohexane — the radicals of the same kind are created.

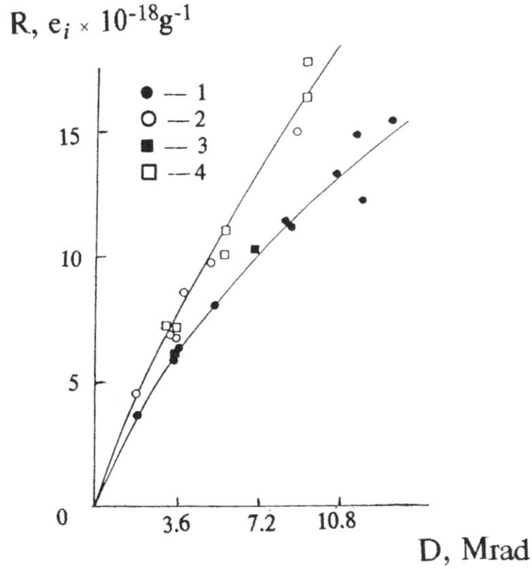

Figure 7.1 Dependence of radical concentration vs. absorbed dose at different regimens of irradiation. 1,3: irradiation of naphthalene and benzene solutions in methylcyclohexane by electrons; 2,4: simultaneous irradiation of naphthalene and benzene solutions in methylcyclohexane by electrons and light.

Now we estimate the magnitude of the ratio $R_{e+h\nu}/R_e$. Substituting in the formula the values of the yields: $G_T = 1.5$ and $G_R = 3.5$, we find $R_{e+h\nu}/R_e = 1.4$. The experimentally obtained magnitude has approximately the same value. The mutual effect of ionizing radiation and light decreases the protection properties of the aromatic compounds with regard to the effect of radiation.

7.3 KINETICS OF RADICAL CREATION UNDER THE ABSORPTION OF LIGHT WITH THE STABILIZED ELECTRONS

Let us consider the kinetics of the radicals creation in case of simultaneous exposure to fast electrons and light of frozen alcohols. It has been mentioned above that under the effect of radiation the stabilization of electrons in the alcohols takes place. These electrons absorbing a quantum of light enter into reaction with alcohol molecules.[4] A dissociation capture takes place resulting in the creation of radicals and the release of electrons from the trap. Then the traps are filled again with electrons and the process is repeated. This is the effect of the simultaneous action of the γ-radiation and light. Such a process can take place continuously during the whole time of simultaneous exposure. Under consecutive exposure because of the absence of free electrons in the time of exposure to light only a single process is possible.

The stabilized electrons are rather long-lived intermediate particles; therefore, to take into account the singularities of simultaneous exposure, they are concealed due to the fact that under consecutive irradiation the increase in the radicals concentration takes place, too.

The concentration of the radicals in different regimens of exposure obeys the inequality $R_{e+hv} > R_{e,hv}$. This inequality takes place when the following conditions are realized:

1. The light quantum energy must be such that the stabilized electron is excited with light to the reaction-capable level.
2. The stationary concentration of the stabilized electrons must not exceed the stationary concentration of the radicals.
3. The irradiation rate must exceed the rate at which the stationary concentration of the stabilized electrons is achieved.
4. The traps of stabilized electrons either can be "utilized" repeatedly or the previously unused ones are filled up, with the dissociation capture of the electron by molecules taking place at each filling-up as in the case of the excitation of the stabilized electron.

To prove the idea of the multiple filling-up of the traps, the measurement for the concentration of the alcohol radicals in ethanol obtained in the different regimes of exposure was undertaken.[9] The measurement showed (Figure 7.2) that the concentration of radicals obeyed the following inequality $R_{e+hv} \geq R_{e,hv} \geq R_e$.

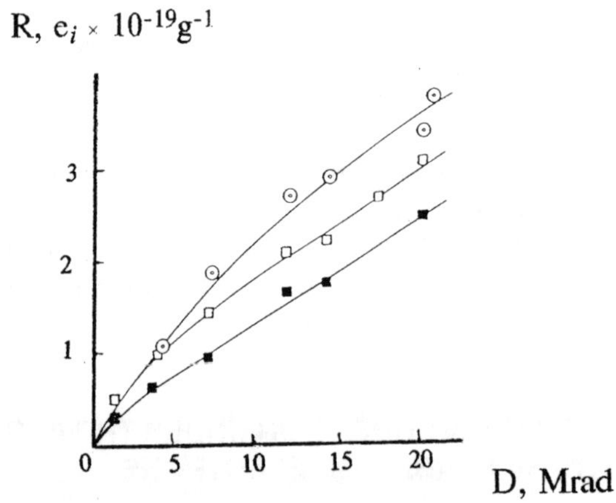

Figure 7.2 Dependence of radicals concentration in ethanol vs. absorbed dose at different regimens of irradiation. ■ = irradiation by ionizing radiation; □ = consecutive irradiation; ⊙ = simultaneous irradiation.

When the irradiation rates are less than those required for establishing the stationary concentration of the stabilized electrons, the ratio is fulfilled: $R_{e+hv} \cong R_{e,hv}$.

The concentration of radicals at different modes of irradiations meets the inequality of high doses: $R_{e+h\nu} > R_{e,h\nu}$. Hence, the experimentally obtained results prove the scheme suggested above.

The photoradiation effect turned out to be somewhat less than the expected one. This seems to be related to the annihilation of radicals under the effect of radiation and not to the violation of condition (4) listed above. This can be elucidated by measuring the concentration of the products in the different regimes of exposure. Another experimental verification[11] of the assumption on the multiple "utilization" of traps (verification of the possibility for continuously carrying out the dissociation capture reaction of the trap-excited electron) is contained in the following.

The investigated substance is exposed to ionization radiation within the time required for establishing the stationary concentration of the stabilized electrons. Then the concentration of the alcohol radicals and stabilized electrons is measured. Further, the sample within the same period of time is exposed to light with $\lambda > 440$ nm and again the concentration of the radicals is determined. After this the substance again is irradiated with ionizing radiation and the stabilized electrons again are excited with light, etc. Such cycles are repeated several times. The irradiation rate was selected so that in each cycle the stationary concentration of the stabilized electrons may be achieved. The measurements showed that in each cycle the electrons are stabilized and under the exposure of the sample to light the electrons disappear and in equivalent number the alcohol radicals appear. This demonstrates that either the traps can be "utilized" multiply or the vacant traps are filled up. Under the above formulated conditions the experiments on the cyclic irradiation of ethanol were set up.

The results of these experiments are shown in Figure 7.3. Several cycles were carried out. In each of them the electrons were stabilized, and under the exposure to light they made a transition into radicals in an equivalent number. This verifies the fulfillment of condition (4).

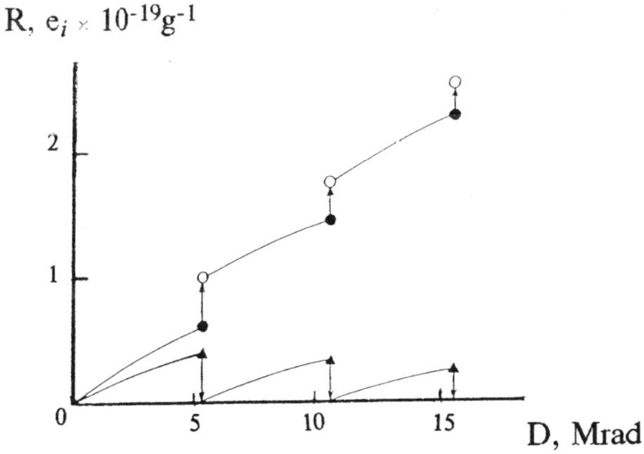

Figure 7.3 Dependence of radical concentration in ethanol vs. absorbed dose at cyclic irradiation. Explanations of figure given in the text. ▲ = concentration of the stabilized electrons; ● = radicals concentration at irradiation by electrons; ○ = radicals concentration at irradiation by light.

Comparing Figures 7.2 and 7.3 it is easy to see that the concentration of the radicals under the simultaneous exposure increases somewhat faster than under the cyclic one. This is explained by the fact that the ionizing radiation rate was selected under the condition of achieving the stationary concentration of the stabilized electrons. Hence, a part of the time the irradiation took place under the filled-up traps; at the same time, under simultaneous exposure, the electrons can stabilize within the whole period of the irradiation time and participate in the creation of the radicals.

7.4 THE ELEMENTARY PROCESSES OF PHOTOCHAIN REACTIONS

As mentioned previously (Section 4.3), different mechanisms of the medium effect on the process of molecular transfer of a particle are possible. In this section, it will be shown how the dynamic properties of the medium can influence the temperature dependence of the rate constant of the radicals' photodissociation which greatly determines the kinetics of photoradical chain reactions.

As is known,[1,2] the free radicals stabilized in the solid matrix absorbing a quantum of light become reaction-capable and can participate in chemical reactions. The development of the photoradical chain, as a rule, takes place according to the following mechanism. Under excitation with light, the radical R_1 can dissociate with the creation of the molecular product M_1 and a new radical R_2 which enters into reaction with the molecules of the substance M. As a result, the product M_2 is created and the initial radical R_1 recovers.

The reaction scheme mechanism has the form

$$\begin{aligned} &R_1 \xrightarrow{\hbar\nu} R_1^*, \\ &R_1^* \rightarrow M_1 + R_2, \\ &R_2 + M \rightarrow M_2 + R_1. \end{aligned} \tag{7.6}$$

The multiple repetition of this sequence of reactions was named the "photoradical chain".[12]

The role of the active center leading the chain is played by the excited radical; i.e., to continue the chain it is necessary for a quantum of light to be absorbed. It is necessary also for the second radical, which is created under the dissociation of the initial radical, to be reaction capable and to enter into reaction with the substance molecules. If this radical is mobile in the given matrix, then due to the reactions of Equation (7.1) the free valences migrate along the substance, as a result of which the radicals meet and the following recombination takes place.

Below, the main attention will be paid to the process of the radicals' photodissociation (the second stage in the reaction scheme of Equation (7.1)); the kinetics of the third stage are also investigated in detail, but we do not dwell upon that here (see Sections 2,4- and 6).

Two mechanisms of the medium affecting the process of the monomolecular decay of particles are possible.[13] The first mechanism is the rearrangement of the medium as a result of the transition and origin of the rearrangement energy. The second mechanism is stipulated by the interaction of the terms between which the transition takes place with the molecular environment. Indeed, since the parameters which determine the interaction of the terms with the medium, corresponding to the initial and final states of the molecule, are different the displacements of these terms under the intermolecular oscillations are different, too. As a result of the molecular oscillations of the radical and molecules interacting with this radical, the mutual location and form of the terms between which the transition takes place change. The displacement of the terms is more the larger amplitude of the thermal oscillations of the medium molecule is. The higher the temperature, the larger the amplitude of the molecular oscillations and, hence, the larger displacements of the terms. Because the matrix element of the transition increases under the approach of the terms, the transition rate constant increases with the increase in the temperature.

In the case of classical intermolecular oscillations, this mechanism leads to the temperature dependence which is described by Equation (4.7). However, one should bear in mind that the parameter α is different for the mono- and bimolecular reactions.

To verify experimentally the kinetic regularities to which the interaction of the terms with the medium leads, experiments were set up and experimental data on the photodissociation of hydrazine and polyvinylene alcohol radicals were gathered.[13] Since, under the extraction of a hydrogen atom, no essential rearrangement of the medium takes place, only the second above-described mechanism determines the temperature dependence of the photodissociation rate constant. Figure 7.4 presents the experimental data on the temperature dependence of the reactions rate constant:

$$N_2H^*_3 \rightarrow N_2H_2 + H \text{ (curve 1)} \tag{7.7a}$$

$$-C_2H_2OH^* - \rightarrow -C_2H_2OH + H \text{ (curve 2)} \tag{7.7b}$$

It is easy to see that the experimentally obtained dependencies agree with Equation (4.7) defined theoretically.

It also follows from theory that the parameter α (see Equation (4.7)) for the monomolecular process is much less than the same parameter for the bimolecular reactions discussed in Section 5.5. Their ratio is approximately equal to the square ratio of the constant of a term coupling with the medium to the constant characterizing the interaction of splitting-off an H atom with the rest of the radical. Comparison of the experimental data on the temperature dependence of the rate constant of the monomolecular (Equation (7.7a)) and bimolecular C_2H_2 + aminoalkyl $\rightarrow C_2H_6$ + $(C_2H_5)_2$ NC_2H_4 reactions obtained in the comparative conditions is conducted in the work.[13] It is easy to see that as expected the experimental values of the parameters α for these reactions differ by the order of magnitude.

For the monomolecular reactions, where the essential conformation transformation of the molecules and the rupture of the coupling with the split-off of large fragments take place, the rearrangement of the medium will play the main role in the formation of the temperature dependence of the rate constant. As far as the discrete final intermolecular states are concerned, the given mechanism leads to the Arrhenius temperature dependence, which is determined by the energy of the medium rearrangement.[13,14] If the final states of the intramolecular subsystem belong to the continuous spectrum, then in case of the classical intermolecular oscillations the rate constant is transformed to the form of Equation (4.7). The value α is expressed through the parameters of the barrier and the energy of the medium rearrangement.[13] To verify experimentally the above dependence of the radical dissociation rate constant on temperature, experiments were set up and the experimental data on the radical photodissociation in the frozen acetic anhydride ($CH_2C(O)OC(O)H_3$) and aminoalkyl ($CH_3CHN(C_2H_5)$) were obtained. In these systems, the alkyl radicals CH_3 and C_2H_5 are the radicals leading the chain. Shown in Figure 7.4 are the results of the experimental study of the temperature dependence of the radical photodissociation rate constant in relative units with the split-off of the radicals CH_3 (curve 3) and C_2H_5 (curve 4) abstraction. According to Equation (4.7) the curves become straight in the coordinates *ln k–T*. Thus, both discussed mechanisms of the medium affect the process of the monomolecular disintegration in the investigated range of temperatures and lead to the linear dependence of the logarithm of the rate constant on temperature, as predicted by theory developed in Reference 13.

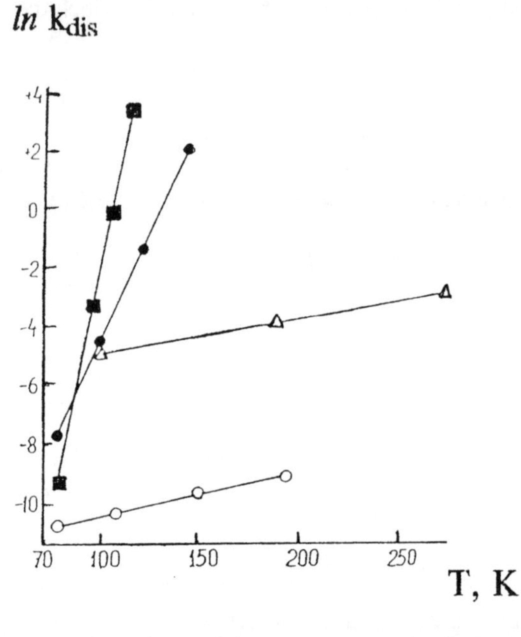

Figure 7.4 Temperature dependence of the rate constant of radicals photodissociation. ○ = atomic hydrogen formation in photodissociation of the hydrazine radicals; △ = atomic hydrogen formation in photodissociation of the polyvinyl alcohol radicals; ● = methyl radical formation in photodissociation of the acetic anhydride radicals; □ = ethyl radical formation in photodissociation of the aminoalkyl radicals.

REFERENCES

1. Shelimov, B.I., Fock, I.B., Voevodskii, V.V., Photochemical destruction of alcohols at low temperatures. Kinetics of methanol destruction, *Kinet. Catalysis,* 4, 539, 1963.
2. Pshezhetskii, S.Ya., Kotov, A.G., Milinchuk, V.K., Tupikov, V.I., Processes of free radicals creation and conversion at consecutive and simultaneous irradiation of γ-radiation and light, *Intl. J. Rad. Phys. Chem.,* 6, 159, 1974.

3. Milinchuk, V.K., Zhdanov, G.S., Pshezhetskii, S.Ya., Formation of free radicals in polymers exposed to the simultaneous action of light and γ-radiation, in *Proc. Third Tihany Symp. Radiat. Chem., Budapest,* 1077, 1972.
4. Smirnov, V.A., Alfimov, M.V., Behaviour of stabilized electrons in alcohol glasses at 77 K, *High Energy Chem.,* 6, 178, 1972.
5. Singh, A., Triplet state formation in pulse radiolysis, *Rad. Res. Rev.,* 4, 1, 1972.
6. Alfimov, M.V., The triplet states in radiation chemistry, *High Energy Chem.,* 6, 3, 1972.
7. Bagdasaryan, Kh.S., Kirjukhin, Yu.I., Sinitsina, Z.A., Biphotonic reactions. General characteristics, *Chem. Phys. Lett.,* 57, 417, 1978.
8. Bagdasaryan, Kh.S., Biphotonic photochemistry, *Sov. Sci. Rev.,* 2, 139, 1980.
9. Kryukov, P.G., Letokhov, V.S., Nikogosyan, D.N., Borodavkin, A.V., Budowsky, E.I., Simukova, N.A., Multiquantum photoreaction of nucleic acid components in aqueous solution by powerful ultraviolet picosecond radiation, *Chem. Phys. Lett.,* 61, 375, 1979.
10. Alfimov, M.V., Buden, V.N., Shamshev, V.N., Radiation yield of molecules in triplet state in solid solutions, *High Energy Chem.,* 6, 188, 1972.
11. Pshezhetskii, S.Ya., Slavinskaya, N.A., Trakhtenberg, L.I., On the mechanism of the effect of simultaneous irradiation by ionizing radiation and light, not absorbable by the substance, *Dokl. Phys. Chem.,* 231, 663, 1976.
12. Chervonenko, V.S., Pshezhetskii, S.Ya., Photoradical chains in γ-irradiated organic substances, *High Energy Chem.,* 9, 559, 1975.
13. Trakhtenberg, L.I., Slavinskaya, N.A., Pshezhetskii, S.Ya., The effect of dynamic properties of the medium on the kinetics of solid-state radical photodissociation processes, *Chem. Phys.,* 134, 127, 1989.
14. Goldanskii, V.I., Trakhtenberg, L.I., Fleurov, V.N., *Tunnelling Phenomena in Chemical Physics,* Gordon and Breach Science, New York, 1989.

8 HETEROGENEOUS RADIATION-CHEMICAL PROCESSES

Chemical processes occurring under the action of radiation in heterogeneous systems are in many cases decisive for most different directions of modern science and technology. Some examples of the practical application of such processes will be given in Chapter 9. Radiation modification of inorganic materials and polymers, development of composite materials, immobilization of biofunctional materials, radiation-corrosion processes, radiation resistance, and modification of semiconductors, radiation heterogeneous catalysis, radiation effect on biological objects — this is by far not a complete list of areas where radiation-chemical heterogeneous processes play a noticeable part.

In spite of their wide prevalence and the significance of such processes for production they have been insufficiently studied. In addition, the initial state of the surface is of great importance. Heterogeneous processes are usually of a multi-stepped character. An important step is transfer of energy, charge, and mass (chemical reagents) over the boundary of media interface. Due to this fact, the mechanism of radiation chemical processes in heterogeneous systems has not been determined in many cases.

This chapter briefly treats some elementary acts of chemical processes occurring on the surfaces of solids under a radiation effect. They include formation of surface defects, surface reactions, and transfer of energy and charge over the surface.

8.1 RADIATION EFFECT ON SURFACE PROPERTIES

Under interaction of radiation with a dispersive solid, different electron-donor, electron-acceptor, and radical surface centers are formed. A charge of the absorption properties of a solid surface at radiation effect is associated with formation of such defects. So, for example, under irradiation of powders of alkaline-earth metals with high-energy electrons, γ-irradiation, or ultraviolet light, surface F_S^+ and V_S^- centers are formed.[1,2] By analogy, with volume defects surface F_S^+ centers are anion (oxygen) vacancies capturing one electron. Anion vacancies may also catch two electrons or be free.[3] V_S^- centers are a single hole located on oxygen ions neighbouring a cation vacancy (absence of metal ion). In modified extrinsic crystals (depending on the kind of atom nearest to the vacancy), F_S and V_S are formed.

As oxygen is absorbed on magnesium oxide containing F_S^+, the ESR signal characteristic of the F_S^+ center rapidly diminishes and the ESR anion-radical oxygen O_2^- signal appears. This proceeds as a result of electron transfer from the F_S^+ center over oxygen and formation of O_2^- which is stabilized on the anion vacancy.[4]

Under absorption of carbon dioxide gas, a transfer of electron proceeds from the F_S^+ center, and the anion-radical $(CO_2)^-$[5] is formed. Observation was also made of anion-radical O_2^- under interaction of oxygen with F_S^0 centers (anion vacancy captured two electrons) of magnesium oxide.[6]

Under recombination of surface F_S and V_S centers, considerable energy (several electronvolts) is released which, in case of transfer on an adsorbed molecule, may lead to a valency bond cleavage. Irradiation of silica whose surface is covered with hydroxyl groups yields in ESR spectra at a temperature 77 K a doublet from hydrogen atoms stabilized on the surface and a central line of complex structures relevant to radiation defects.[8,9] There were observed both electron (F_S centers)[10,11] and hole[12] vacancy centers. Formation of stabilized H atoms and strong surface dehydroxylation at the interaction of silica gel with radiation appeared around the creation of radical centers (Si· and SiO·). Moreover, in studying adsorption of different gases on the surface of silica gel, diamagnetic centers have been discovered: sililene groups (two coordinated silicon atoms >Si:), syladioxyrane groupings (> Si$\langle$O–O$\rangle$), etc.[13,14]

Thus, even a brief consideration of heterogeneous radiation chemical processes shows that the adsorption properties of dispersive solids change greatly as a result of the action of ionizing radiation. This is associated with formation of different surface defects. More details about the properties of surface centers in semiconductors and dielectrics can be found in Reference 15.

8.2 ACTIVE PARTICLE REACTIONS

Irradiation of a solid brings about creation of chemically active particles, not only in the volume but also on the surface. These particles may react with both neighbouring molecules and molecules from the gaseous phase. In any case, the reactivity of these particles will be related to the state of the molecular environment. This section considers specific features of chemical transformations proceeding on a solid surface.

8.2.1 Elementary Physical and Chemical Processes on the Surface

There are several physical and chemical processes in which an active particle on the surface may participate. First of all, there are chemical reactions with radicals, ions, and molecules from the nearest environment. The difference in reactions in volume is the geometry of the reaction complex, as well as the large amplitude of active particle displacement from the equilibrium state. Since the rate constant of the solid-phase chemical reaction depends on the frequency of molecular vibrations (see Chapters 4 to 6), an essential influence may be caused by temperature dependence of this frequency. In the region of phase transitions, the frequencies of intermolecular vibrations depend very much on temperature. At the same time, the temperatures of phase transitions in volume and on the surface are different. Consequently, noticeable accelerations of volume and surface chemical transformations must proceed in different temperature regions.

If the environment of the active particle does possess reactivity, then a search for a partner is possible as a result of diffusion. The effectiveness of such chemical processes will be under the control of active particle concentration and migration rate. The temperature and concentration dependence of the diffusion coefficient was studied mainly in the experiments. See, for example, Reference 16, where a rather large volume of experimental data is presented. In this work, the atomic diffusion of hydrogen, deuterium, and tritium on the surface of tungsten was studied. A noticeable isotope effect is discovered which cannot be explained only by the difference of zero oscillation frequencies of these atoms. In addition, the low-temperature limit of the diffusion coefficient is observed which, along with the isotope effect, indicates the tunneling mechanism of atomic transfer over the surface.

Another possibility of the chemical transformation of an adsorbed particle is to react with molecules of the gaseous phase (the case of surface bombardment by active particles is explained below) or to leave the surface and transit in the volume. Such a process of field desorbtion of boron ions was studied, for example, in Reference 17. The range of temperatures in which the studied were carried out was 1 to 240 K. The transition from the low-temperature plateau to the temperature-dependent part of the rate constant proceeds in the region of 100 K. This fact demonstrates that the boron atom does not make a transition through the traditional activation mechanism. In order to make a choice in favour of some desorbtion mechanism, it is necessary to get information on the potential barrier as well as on the change of the barrier under the action of the electrical field. Nevertheless, among the results of the experiment there are no data which would contradict the tunneling mechanism.

8.2.2 Exposure of Organic Molecules Adsorbed on a Surface to an $O(^3P)$ Beam

Bombardment of surfaces covered with organized organic thin films (OOTF) of amphiphilic molecules with an atomic beam of ground state oxygen $O(^3P)$ has been described.[18,19] OOTFs were prepared on glass and Si substrates using both Langmuir-Blodgett (LB)[20] and self assembly (SA)[21] techniques. Three types of LB films (cadmium arachidate, cadmium behenate, and deuterated cadmium behenate) and two types of SA films (octadecyltrichlorosilane and methyltrichlorosilane) were studied. The experiments were done under high-vacuum conditions. The sample temperature was held constant during the exposure. The range of temperatures was 240 to 400 K. The beam of $O(^3P)$ atoms was produced by a 2450-MHz microwave discharge. The exposure resulted in chemical changes in OOTF. They were detected by FTIR and XPS.

The main path of the reaction of the films with atomic oxygen was found to be[18,19]

$$\text{C–H} + \text{O} \rightarrow \text{C–O–H}.$$

The insertion of the oxygen atom into the CH bond is typical for reactions of electronically excited oxygen, $O(^1D)$, with hydrocarbons.[22] The reaction of saturated hydrocarbons with $O(^3P)$ in the gas phase occurs by abstraction mechanism:[23]

$$\text{C–H} + \text{O} \rightarrow \text{C}\cdot + \cdot\text{OH}.$$

Kinetics measurements showed that the reactions of both methyl and methylene groups with oxygen are of a pseudo first order. The kinetics of CH_2 depletion was found to be polychronic, probably because of an inequality of reactive sites due to the chains' conformations.[18]

The reaction probability for the investigated process was calculated.[18,19] The obtained value was several orders of magnitude larger than the appropriate value for reaction of the gas phase homologs.[24] The activation energy for the reaction on the surface was substantially less than in the gas phase. The kinetic energy of oxygen atoms did not affect their ability to react. The apparent activation energy in a temperature range of 273 to 333 K was found to be about 0.7 kcal/mol.[18]

Reactivities of the methyl groups in "short" methyltrichlorosilane and "long" octadecyltrichlorosilane chains were compared. A depletion of the methyltrichlorosilane methyls was about five times slower than that of octadecyltrichlorosilane ones.[18]

The model, which explained the above results, was based on the assumption that a collision complex C–H···O is stabilized due to energy exchange with low frequency vibration modes of the adsorbed chains.[18,19] The probability of finding the collision complex in the bent configuration increases with the lengthening of its lifetime. The bent configuration is the most preferable for the reaction, because in this configuration the potential energy surfaces of $O(^3P)$ and $O(^1D)$ cross. The states mix, and $O(^1D)$ reacts with hydrocarbons without a potential barrier.[22] The longer the chain, the wider the range of the vibration modes' frequencies and the more effective the stabilization of the collision complex. The longer the lifetime of the collision complex, the greater the probability of finding it in the reactive configuration. Therefore, reactivity of a monolayer increases with the length of its chains.

In addition, the protective properties of the LB multilayer films were studied.[18] Temperature of the film is an essential factor which determines their protective properties. At 300 K the upper layer protects the inner layers from destruction by bombardment with atomic oxygen. However, at higher temperatures the phase transition from the orthorhombic to hexagonal structure occurs,[25] the chains start to rotate freely, and a depletion corresponding to more than one layer was observed.[18]

It should be noted that in the works discussed in Reference 16, significant acceleration of the chemical process in the phase transition region is reported. It is not surprising, since in the phase transition region the frequency of intermolecular vibrations sharply decreases and the amplitude increases. This leads to a noticeable increase of the rate constants of the corresponding chemical process (see Section 5.3).

8.3 RADIOLYSIS OF SUBSTANCES IN THE ABSORBED STATE

The regularities of radiation chemical transformations differ greatly for heterogeneous and homogeneous systems. It should be pointed out that a significant increase of the radiation product appears in heterogeneous conditions as compared to homogeneous ones. This phenomenon is observed in different systems and is a convincing argument for the absorbed energy redistribution between a solid and adsorbed molecules. So, for example, in radiolysis of a heterogeneous system, *n*-pentan-silica gel hydrogen yield was sufficiently higher than in radiolysis of liquid pentane.[26] Similar results were obtained under γ-irradiation of nitrogen dioxide molecules absorbed on silica gel,[27] radiolysis of methanol adsorbed on different solid surfaces (insulators, semiconductors),[28] and irradiation of isopropylbenzene-aluminosilicate.[29]

In some cases, the radiation chemical process yields in heterogeneous systems reach very high values. Under radiation oxidation of 70 different hydrocarbons on silica gel, aluminium oxide, and other oxides, the radiation yield of oxide products was larger by as much as two orders of magnitude.[30,31] Irradiation of a nitrogen-hydrogen mixture at atmospheric pressure on aluminium oxide led to formation of ammonia with a radiation yield of up to 500 molecules per 100 eV.[32]

Chain radiation-catalytic processes in heterogeneous systems are a good example illustrating the necessity of accounting for such processes as creation of surface defects, reactions of surface intermediate active particles, and energy and charge transfer over the boundary of the media interface in hetergeneous systems. It should be noted that the effect of energy transfer is constrained mainly by the first layer of the adsorbed molecules.[33] It is clear that the above-mentioned regularities are created by the transfer of energy absorbed by the substance over the surface. Various aspects of this phenomenon were discussed in works cited below.

Under action of light and high-energy irradiation on adsorbent-adsorbate systems, a formation of ions and electron-excited atoms and molecules and their emission in the gas phase was observed. The formation of electron-excited atoms and molecules as a result of energy transfer from a solid to adsorbed substance proceeds due to different processes.

Under light exposure of different adsorbed dyes[34] and metalporphyrins[35] and as a result of triplet-triplet energy transfer over the absorbed oxygen, an electron-excited oxygen molecule, singlet oxygen, is formed. A process of triplet-triplet energy transfer from light-excited solid organic substances (naphthalene, 9,10-anthraquinone, etc.) to gas phase molecules (1,3-pentadiene)[36] was observed. The interaction of oxygen molecules with the triplet excitons of solid films of tetracene,[37] Bengal rose,[38] and trypaflavine[38] causes formation of singlet oxygen. Electron-excited molecules are also formed at the energy transfer from the excited surface center containing ions of transition metals (for example, molybdenum and vanadium) to an adsorbed molecule.[40] In the process of extinguishing the phosphorescence of the surface center by oxygen, a singlet oxygen arises.

The source of energy, transferred by a solid, and excitation of adsorbed molecules may be annihilation of defects in the near-surface layers of solids and structure-chemical processes (phase transformations of a solid, reaction of surface reduction of some oxides). Annealing of quartz near-surface defects (created as a result of a mechanical destruction or chemical reaction) and those of nonstoichiometric oxide $Co_{3-x}O_4$ in oxygen atmosphere leads to formation of electron-excited oxygen molecules (singlet oxygen).[41] The phase conversion of Magnelli in nonstoichiometric oxide-vanadium pentoxide upon heating of the oxygen atmosphere is the cause of singlet oxygen formation.[42] In a set of oxides, in the surface atoms which are bound rather weakly with the crystal lattice, a surface reduction is possible under interaction with oxygen atoms. Surface reduction, for example of mercury oxide, is accompanied by a generation of singlet oxygen.[43]

The regularities of formation and emission of different ions and electron-excited molecules upon bombardment of solid surfaces (of metals and their oxides) by electrons and ions of different gases are treated in Reference 44.

Energy transfer also proceeds from adsorbate to absorbent. Spectral sensibilization of photoconductivity of solids by surface-absorbed dyes and heterogeneous deactivation of electron-excited atoms and molecules of the gas phase were observed.[45] Singlet oxygen deactivation in the surface of zinc oxide as a result of the energy transfer of singlet oxygen leads to the ionization of small traps of zinc.[46] In other cases, a multi-phonon process of singlet oxygen relaxation on the quartz surface and KBr are apparently realized.[47,48] It is indicated by the slight effectiveness of annihilation and considerable lifetime (10^{-2} to 10^{-3} sec) of singlet oxygen in the surface. In absorption of singlet oxygen on metals, a mechanism of energy relaxation on conduction electrons was suggested. Nevertheless, the main channel of deactivation due to its high reactivity is considered to be hemosorbtion of singlet oxygen.[47,60] Chemically, in active gases (for example, inert gases), only the transfer of energy is observed. For instance, it was shown that under interaction of metastable xenon atoms with a solid xenon surface, excitons are formed in a solid.[51]

There are different points of view about the energy transfer mechanism of electron excitation. For example, on the basis of the radiolysis regularities of adsorbed methanol and heavy cyclohexane a conclusion is made that the energy transfer is related to recombination of the hole, captured by an adsorbed molecule, with an electron of the conductivity band. These conduction electrons are brought about under the radiation effect.[28,52] If the width of the forbidden band is larger than the dissociation energy of the adsorbed molecule, then the sensibilization of decomposition by adsorbent is effective, otherwise sensibilization is not observed.

The authors of Reference 53 hold a somewhat different point of view. The energy transfer over the adsorbed molecule, to their mind, proceeds as a result of electron recombination with a hole localized in the defect. Such a hole is connected with donor-acceptor and exchange interaction with the adsorbed molecule which is characteristic of molecules possessing donor properties. In this case, the effectiveness of energy transfer is determined mainly by the depth of the electron and hole traps and the adsorption heat of molecules rather than width of the forbidden band. For acceptor molecules, a dissociative capture of free or localized electrons by adsorbed molecule is possible — for example, in radiolysis of ammonia on zeolites.

Of some interest are the results of radiolysis of the hydroxyl cover of silica gel.[54] It was found that the energy necessary for OH bond dissociation is transferred from the recombination level corresponding to the surface defect. By increasing or decreasing the number of recombination centers by different techniques (for example, by heating or adsorption of molecules), it is possible to either increase or decrease the radiation yield of the dissociation of hydroxyl groups.

A recharge mechanism of adsorbed molecules or surface defects has been suggested.[55] The electrons and holes were considered there as free valences of the surface (Frenkel excitons also may play part in free valences). If a free valence is localized on the adsorbed molecule, then valence bond cleavage of the molecule takes place and the valence forms of hemosorption are produced. In the case of electron and hole localization in surface defects, an adsorption center is formed which plays a part in a free valence. A relationship between catalytic activity of irradiated silica gel and concentration of adsorbed centers (extrinsic aluminium captured a nonequilibrium hole) has been discovered also.[56]

The dissociation of water on semiconductor surfaces was treated on the basis of the donor-acceptor mechanism of hemosorption.[15] A donor-acceptor complex arising on the surface (the coordination-unsaturated atoms of the surface) may act as a trapping center of nonequilibrium charge carriers, and the localization of the charge carriers may be realized by a neighbouring defect. The capture of a photo-initiated hole in adsorbed slow states leads to additional deformation of water molecule and its dissociation.

Similar mechanisms were suggested in discussing other processes. For instance, very high radiation chemical process yields of initiating surface polymerization (e.g., SiO_2, MgO, Al_2O_3) are associated with the recombination of excess charge carriers, stabilized in the surface, with proton participation of surface hydroxyl group formed as a result of irradiation and some other reasons.[57,58]

Even such a brief review shows that in the photolysis and radiolysis of adsorbed molecules, the energy transfer from adsorbent to adsorbate has a different cause, brings about a large yield of products, and reduces the threshold of dissociation.

REFERENCES

1. Abraham, M.M., Chen, Y., Uuruch, W.P., Formation and stability of V^- and V_{Al} centers, *Phys. Rev. B.,* 9, 1842, 1974.
2. Lunsford, J.H., A study of irradiation-induced active sites on magnesium oxide using electron paramagnetic resonance, *J. Phys. Chem.,* 68, 2312, 1964.
3. Henderson, B., Wertz, J.E., *Defects in the Alkaline Earth Oxides,* Taylor and Francis, London, 1977.
4. Lundsford, J.H., ESR of absorbed oxygen species, *Cat. Rev.,* 8, 135, 1973.
5. Ben Taarit, Y., Vedrine, J.S., Naccacke, C., de Montgolfier, Ph., Coadsorption of CO_2 and O_2 on MgO. An experimental and theoretical ESR study, *J. Chem. Phys.,* 2880, 67, 1977.
6. Grigoriev, E.I., Trakhtenberg, L.I., Tchasov, V.V., Shets, V.A., Nitroxyl radical interaction with electron-donor sites on oxide surfaces, *Russ. J. Phys. Chem.,* 68, 1340, 1994.
7. Kassin, H.A., Matthew, J.A.D., Green, B., Electronic structure of the surface F centre in ionic crystals, *Surf. Sci.,* 74, 109, 1978.
8. Kazansky, V.P., Pariisky, G.B., Voevodsky, V.V., Radiated-induced processes on the silica gel surface, *Disc. Farad. Soc.,* 31, 203, 1961.
9. Muha, G.M., Electron spin resonance studies of aromatic hydrocarbons absorbed on silica-alumina II anthracene, *J. Phys. Chem.,* 71, 640, 1967.
10. Jamada, J., Hasegawa, A., Miura, M., Electron spin resonance studies of formation of anion radicals on γ-irradiated silica gel, *Bull Chem. Soc. Jap.,* 42, 1836, 1969.
11. Wong, P.K., Willard, J.E., Reactions of carbon dioxide, sulfur dioxide, nitric oxide, and alkyl halides with γ-irradiated silica gel, *J. Phys. Chem.,* 73, 2226, 1969.
12. Feigl, F.J., Fowler, W.B., Yip, K.L., Oxygen vacancy model for the E'_1 center in SiO_2, *Sol. State Comm.,* 14, 225, 1974.
13. Radtsig, V.A., "Reactive silica" — new, *Khimicheskay Fizika,* 10, 1262, 1991.
14. Morterra, C., Low, M.J.D., Reactive silica: novel aspects of the chemistry of silica surfaces, *Ann. N.Y. Acad. Sci.,* 220, 135, 1973.

15. Kisilev, V.F., Krylov, O.V., *Electronic Phenomena in Adsorption and Caralysis on Semiconductors and Dielectrics,* Vol. 8, Springer-Verlag, Berlin, 1988.
16. Wang, S.C., Gomer, R., Diffusion of hydrogen, deuterium, and tritium on the (110) plane of tungsten, *J. Chem. Phys.,* 83, 4193, 1985.
17. Menad, A., Kingham, D.R., Evidence for the quantum mechanical tunnelling of boron ions, *J. Phys. C: Solid State Phys.,* 18, 4539, 1985.
18. Paz, Y., Trakhtenberg, S., Naaman, R., Reaction between $O(^3P)$ and organized organic thin films, *J. Phys. Chem.,* 98, 13517, 1994.
19. Paz, Y., Trakhtenberg, S., Naaman, R., New evidence for surface nonspecific catalysis: the reaction between organized organic thin films and $O(^3P)$, *J. Amer. Chem. Phys.,* 116, 10344, 1994.
20. Blodgett, K.B., Films built by depositing successive monomolecular layers on a solid surface, *J. Amer. Chem. Soc.,* 57, 1007, 1935.
21. Sagiv, J., Organized monolayers by adsorption. 1. Formation and structure of oleophobic mixed monolayers on solid surfaces, *J. Amer. Chem. Soc.,* 102, 92, 1980.
22. Luntz, A.C., Chemical dynamics of the reactions of O(1D) with saturated hydrocarbons, *J. Chem. Phys.,* 73, 1143, 1980.
23. Andersen, P., Luntz, A.C., The chemical dynamic of the reactions of $O(^3P)$ with saturated hydrocarbons, *J. Chem. Phys.,* 72, 5842, 1980.
24. Herron, J.T., Evaluated chemical kinetic data for the reactions of atomic oxygen $O(^3P)$ with saturated organic compounds in the gas phase, *J. Phys. Chem. Ref. Data,* 17, 967, 1988.
25. Tippmann-Krayer, P., Kenn, R.M., Mohwald, H., Thickness and temperature dependent structure of Cd arachidate Langmuir-Blodgett films, *Thin Solid Films,* 210, 577, 1992.
26. Caffrey, J., Allen, A., Radiolysis of pentane adsorbed on mineral solids, *J. Phys. Chem.,* 62, 33, 1958.
27. Sutherland, J., Sthi, D., Goodrich, R., Radiolysis of carbon dioxide in the absorbed state, *J. Phys. Chem.,* 71, 4422, 1967.
28. Zhabrava, G.M., Vladimiriva, V.I., The main regularity and role of electronic factors in catalytic processes which follow under action of ionizing radiation, *Russ. Chem. Rev.,* 38(4), 711–739, 1969.
29. Hentz, R., Radiation-induced reactions of isopropylbenzene on silica-alumina, *J. Phys. Chem.,* 68, 2889, 1964.
30. Kohn, H.W., Surface carbonium ions produced by irradiating silica gel, *J. Phys. Chem.,* 66, 1185, 1962.
31. Kohn, H.W., Oxidation of hydrocarbons absorbed on oxide catalyst induced by cobalt-60 γ-rays, *J. Phys. Chem.,* 68, 3129, 1964.
32. Coekelbergs, R., Crucq, A., Frennet, A., *Advances in Catalysis and Related Subjects,* Vol. 13, Academic Press, New York, 1962, 55.
33. Sutherland, J., Allen, A., Radiolysis of pentane in the absorbed state, *J. Amer. Chem. Soc.,* 83, 1040, 1961.
34. Kearns, D.R., Physical and chemical properties of singlet molecular oxyden, *Chem. Rev.,* 5, 395, 1971.
35. Twarowski, A.I., Good, L., Phosphorescence quenching by molecular oxygen: zinc tetraphenylporphin on solid supports, *J. Phys. Chem.,* 91, 5252, 1987.
36. Daubendick, R.L., Magid, M., McMilan, G.K., Electronic energy transfer at the gas-solid interface, *Chem. Commun.,* 2, 218, 1968.
37. Frankevich, E.L., Sokolic, I.À., Photooxidation of solid tetrazene. Influence of magnetic field, *High Energy Chem.,* 6(5), 433–439, 1972.
38. Eisenberg, W.E., Taylor, K., Veltman, J., Generations of $O^2(' \Delta_2)$ by direct absorption of radiation by groundstate oxygen, *J. Amer. Chem. Soc.,* 104, 1104, 1982.

39. Grigoriev, E.I., Myasnikov, I.A., Tsivenko, V.I., Photo-stimulated formation of singlet oxygen in heterogeneous systems: dye-solid-gas, *Russ. J. Phys. Chem.,* 55, 2907, 1981.
40. Yakunichev, M.V., Myasnikov, I.A., Tsivenko, V.I., Photo-stimulated formation and emission of singlet oxygen from the surface of quartz doped with molybdenum ions, *Russ. J. Phys. Chem.,* 59, 756, 1985.
41. Zav'yalov, S.A., Myasnikov, I.A., The role of certain structural-chemical transformations of solids in heterogeneous processes, *Russ. J. Phys. Chem.,* 62, 2786, 1988.
42. Zav'yalov, S.A., Myasnikov, I.A., Zav'jalova, L.M., The role of structural-chemical transformations at the surface of solids in the formation of singlet oxygen and its participation in naphthalene catalytic oxidation, *Dokl. Phys. Chem.,* 284, 378, 1985.
43. Whilow, S.H., Findlay, F.O., Single and double electronic transitions in molecular oxyden, *Canad. J. Chem.,* 45, 2087, 1967.
44. Thomas, E.W., Formation of excited states by ion impact on surface, *Prog. Surface Sci.,* 10, 383, 1980.
45. Akimov, A.Z., Cherkasov, Y.A., Cherkashin, M.I., *Sensibilised Photoeffect,* Nauka, Moscow, 1980 (in Russian).
46. Myasnikov, I.A., Zav'yalov, S.A., Grigoriev, E.I., et. al, On interaction of electron-excited particles with surface of solid, *Russ. J. Phys. Chem.,* 55, 1840, 1981.
47. Grigoriev, E.I., Myasnikov, I.A., Tsivenko, V.I., Deactivation of singlet oxyden on glass and zinc oxide surface, *Russ. J. Phys. Chem.,* 57, 312, 1983.
48. Solonitsyn, Yu.P., Basov, L.L., Ryabchuk, V.K., Photo-sorbtion and photo-catalytic properties of oxide and alkali halide crystals, *Russ. J. Phys. Chem.,* 54, 1494, 1980.
49. Ryskin, M.E., Shub, B.R., Heterogeneous relaxation of singlet oxyfen, *React. Kinetics Catalysis Lett.,* 17, 71, 1981.
50. Ryskin, M.E., Shub, B.R., Temperature dependences of heterogeneous relaxation rate of singlet oxygen on quartz and platinum, *Russ, J. Phys. Chem.,* 1, 212, 1982.
51. Grigoriev, E.I., Kupriyanov, L.Yu., Myasnikov, I.A., Trakhtenberg, L.I., Deactivation of metastable xenon atoms on a solid xenon surface, *Russ. J. Phys. Chem.,* 60, 467, 1986.
52. Sagert, N.H., Reid, J.A., Robinson, R.M., Radiolysis in the absorbed state. III. Methyl iodide absorbed on silica gel, *Canad. J. Chem.,* 48, 17, 1970.
53. Psheezhetskii, S.Ya., Ed., *EPR Free Radicals in Radiation Chemistry,* Khimia, Moscow, 1972 (in Russian).
54. Unpublished results of Grigoriev, E.I., Trakhtenberg, L.I., Vorontsov, P.S.
55. Baru, V.G., Volkenstein, F.F., *Irradiated Effect on Surface Properties of Semiconductors,* Nauka, Moscow, 1978 (in Russian).
56. Boreskov, G.K., Kazanski, V.B., Mishenko, Iu.A., Pariskii, G.B., The nature of active centers in the reaction involving the isotopic exchange of hydrogen on γ-rayed silica gel, *Dokl. Phys. Chem.,* 157, 384, 1964.
57. Mezhirova, Z.P., Sheinker, A.P., Abkin, A.D., The radiation polymerization of acryl-onytryl and methylmetacrylate as influenced by semiconducting additives, *Docl. Phys. Chem.,* 153, 1378, 1963.
58. Bruk, M.A., Radiation polymerization of monomers adsorbed on surface of solids, *Russ. Chem. Rev.,* 56, 148, 1987.

9 APPLIED PROBLEMS OF RADIATION CHEMISTRY OF SOLIDS

The previous chapters in this book have discussed mechanisms and kinetics of different radiation chemical, solid-phase processes. The material of this chapter is associated with application of these issues. Papers devoted to applications of radiation chemistry, as a rule, do not treat correlation with reaction mechanisms. However, such a correlation exists and can be established. Here, an attempt is made in a concise form to cover rather fully applied problems where radiation plays an important part. Both traditional issues already described in other monographs[1] devoted to various aspects of radiation chemistry applications and relatively new problems are discussed. They include high-temperature superconductivity, fullerenes, and the effect of small radiation dose rates.

Furthermore, a new original technique of utilizing radioactive wastes is presented and much attention is devoted to microelectronics. Apart from discussing various problems of solid-phase radiation chemistry, consideration is given to biopolymers, as well. This is not done by accident, since giant molecules, owing to their structure, develop a number of properties characteristic of the solid phase.

9.1 RADIATION MODIFICATION OF MATERIALS

Radiation modification of materials is used widely to obtain materials with relevant operational properties. At present radiation chemical engineering may be said to be a new field of science. Use of ionizing radiation allows not only the development of unique materials or materials of improved quality but also in some cases reduced costs of raw material and energy together with reduction of industrial wastes.

9.1.1 Modification of Inorganic Materials

It is known that inorganic materials (metals, alloys, ceramics, glasses, etc.) possess, as a rule, high radiation resistance.[2] A change of operational properties in these materials under interaction with radiation takes place mainly due to defect formation. For this reason, although the possibilities of radiation modification of these materials are limited due to a near-surface area, it suffices practically for a large number of promising applications.

9.1.1.1 Metals, Alloys

The surface of inorganic materials is mostly sensitive to radiation effect. A powerful electron or ion beam of high energy (~1 MeV) can be used to modify material surfaces. As a result, as surface hardness increases, surface irregularities are smoothed out, corrosion resistance improves, and the surface becomes refined. Such results are achieved by means of implantation of relevant elements in a near-surface area (1 to 20 μm), as well as by melting or evaporating of the near-surface material area with further, rapid cooling. It occurs due to the local character of beam energy absorption and high temperature formation in the surface. It becomes possible to create amorphous layers of matter and other nonequilibrium microstructures. For example, a surface treatment of cutting tools by electron and ion beams increases surface hardness by 2 to 5 times and fatigue and abrasive strength by 1.8 to 8 times.[3-7] Use of a carbon-proton beam has led to reducing aluminum surface irregularities of ceramics and magnetic alloy, increasing steel surface hardness by 3 times, and decreasing titanium surface irregularities by 10 times.[8]

A difference in the effectiveness of high-energy radiation absorption allows different ceramic surfaces to be connected.[9] The reason lies in the dependence of radiation absorption coefficient on the effective atomic number of material. In this case, as expected, the energy of ionizing radiation is absorbed in the basic material with a higher atomic number (see Chapter 2). As a consequence, only partial sample melting takes place which allows development of new composite materials.

To increase corrosion resistance of metals and alloys, the ion implantation mechanism can be used.[10] Incapsulation of extrinsic atoms in semiconducting substances is possible by both implantation[11] and radiation-induced nuclear reactions[12] (radiation alloying). As an illustration, one can use alloying of silicon by phosphor. Under silicon irradiation with slow neutrons, an unstable silicon isotope is formed which then transforms into a stable isotope of phosphor; the resultant atoms of phosphor are distributed uniformly in silicon. Nuclear reactions are also used to produce tritium. Radiation of lithium-doped silica with slow neutrons leads to formation of tritium.[13]

Solid-phase synthesis of inorganic materials by use of powerful electron beams has some advantages compared to conventional methods of ceramic materials synthesis. For example, these techniques are used to synthesize high-temperature superconducting ceramics,[14] and to produce magnetic soft materials (e.g., lithium-titanium ferrites) and BeO-Fe_2O_3 systems.[15] The radiation method of synthesis is promising for developing materials out of elements with high melting temperatures because under radiation effect temperatures required for synthesis are reached rather rapidly. Synthesis of superconducting high-temperature ceramics in the field of intensive Co^{60} γ-radiation also leads to improving its properties.[16] It was shown that an ammonia synthesis catalyst synthesized by means of a powerful fast electron beam possesses higher (~20%) activity and slower speed of degradation.[17] The examples given show how important the application of radiation modification in inorganic materials is.

9.1.1.2 High-Temperature Superconductors

A phenomenon of high-temperature superconductivity was discovered in 1986.[18] It was shown that metaloxyde ceramic compounds become superconducting (electronic resistance is equal to zero) at temperatures close to the temperature of liquid nitrogen boiling (77 K) and higher. It is obvious that high-temperature superconductors may have wide application in practice. These materials can be applied in devices operating in radiation conditions (ultraviolet, ionizing radiation) and, consequently, it is necessary to study their radiation resistance.

Under interaction of high-energy radiation with high-temperature superconductors, the temperature of transfer into a superconducting state (T_c) decreases, the width of a the transition temperature range (ΔT_c) increases, and the critical current of a superconducting state failure and volume of superconducting phase decrease. These changes in superconductor properties are usually associated with formation of defects (atomic displacement). The magnitude of displacement does not depend on the type of radiation (electrons, protons, heavy ions) and is determined only by the number of defects produced.[19] The formation of defects in this case, as a rule, occurs as a result of the elastic interaction of a high-energy particle with sample (see Chapter 2). So, for example, irradiation of $YBa_2Cu_3O_{7-\delta}$ ceramics by protons with energy over 1 MeV leads to decreasing T_c by 9 K and decreasing critical current by 30 times at the fluence 10^{19} cm^{-2}.[20] Under further irradiation the superconducting properties of samples are destroyed. Similar regularities were observed under oxygen ion irradiation of ceramics consisting of $YBa_2Cu_3O_{7-\delta}$ and Bi–Pb–Sr–Ca–Cu–O.[21]

The fundamental cause of the changes observed during interaction with radiation is the knockout of oxygen form Cu–O chains with which the effect of superconductivity is associated. The removal of oxygen and transition of the lattice $YBa_2Cu_3O_{7-\delta}$ from the orthorhombic (superconducting phase) to tetragonal (nonsuperconducting phase) happens at (displace per atom) dpa = 0.1. With an increasing number of defects (dpa = 0.3) the phase Cu_2O and the mixture of barium and yttrium amorphous oxides appear.[22] It is also assumed that the radiation damage of the grain boundaries of the superconducting phase plays an important part in the degradation of superconducting properties.[23] When irradiating ceramics with X-ray and γ-radiation, their superconducting properties degenerate, too. It should be pointed out that in this case a formation of defects proceeds through a subthreshold mechanism. So, under γ-irradiation of modified (replacement of copper atoms by iron) ceramic $YBa_2Cu_3O_{7-\delta}$, a decrease in temperature of transfer to a superconducting state, T_c, at dose rates ~3 MGy was observed.[24] It was associated with primary removal of oxygen from the boundaries of superconducting domains. An increasing degree of sample alloying by iron reduced the radiation resistance of ceramics.

X-ray irradiation (K_α 20.2 keV and K_β 22.7 keV) of ceramic $ErBa_2Cu_3O_{7-x}$ also led to suppressing its superconducting properties. The reason for this, according to Reference 25, is oxygen removal from the Cu–O chain. On the other hand, according to Reference 26, changes in superconducting properties of $GdBa_2Cu_3O_{7-x}$ under γ-radiation with a dose rate of 180 MGy are not discovered. Evidently, the radiation resistance of superconducting ceramics in this case depends strongly on "biography"-specific features (defects, impurities, state and composition of grain boundaries of superconducting domains, etc.).

9.1.1.3 Fullerenes

Fullerenes are a new form of carbon along with the existing forms in the nature such as graphite and diamond. They were discovered in studying a laser-induced evaporation of graphite in 1985.[27] One of the representatives of the fullerenes family, C_{60}, represents a spherical molecule consisting of 20 hexagons and 12 pentagons bound in such a manner that they form a volume molecule resembling a soccer ball in shape. Such a form of carbon possesses a number of unique properties. These substances have numerous potential applications, for example, in pharmacology, catalysis, development of new dyes, composites, carbon fibers, polymers, optical materials, etc.

One of the techniques for modifying fullerenes is irradiation with high-energy and ultraviolet radiation. Interaction of C_{60} with fast charged particles (*Xe* ions with the energy of 320 keV,[28] protons with the energy of 1.3 MeV, and iodine ions with the energy of 55 MeV[29]) leads to an effective destruction of the fullerene and its transformation into amorphous carbon. Apart from this, the structural changes of a fullerene molecule were observed, bringing about its oxidation under interaction with oxygen.[29] Irradiation of C_{60} solid films with argon laser visible light gave rise to polymerization of fullerene molecules.[30] A suggestion was made to use a mechanism of 2 + 2 cycloaddition for producing C_{60} clusters. The number of molecules in such a cluster amounted to 10. Apart from this, polymerization is a consequence of C_{60} film irradiation with X-rays with energy of 1487 eV.

Under pulsed radiolysis of fullerenes in different solutions, formation of anion and cation radicals of C_{60} and radical adducts of C_{60}–R[31] takes place. In particular, γ-irradiation of C_{60} powder leads to appearance in the ESR spectra of a symmetric singlet signal with g-factor 2.0030 and 1.4 Gauss width (9 GH).[32] The behavior of the ESR signal in oxygen bleeding differs from the behavior of signals with the same parameters which are characteristic of insufficiently pure samples that were observed before irradiation. If the ESR signal of an unirradiated sample under oxygen bleeding increases, which was observed, for example, in Reference 33, in an irradiated sample under oxygen bleeding the ESR signal decreases. In both cases, the changes are reversible. The ESR signal of irradiated C_{60} fullerene behaves in the same manner as paramagnetic centers behave in polymers with a system of double bonds.[34] The nature of such centers has not been fully explained. It is assumed that paramagnetic centers in such systems are formed as a result of deformation of separate molecules at the expense of intramolecular interactions. It should be noted that a molecule C_{60} (triplet level 37.5 kcal/mol) is the effective generator of singlet oxygen and may cause a strong photocatalytic effect on biological systems.[35]

The size of the inner cavity of fullerenes (~10 Å) allows different atoms, e.g., metals, to be implanted into it. This is how endohedral metallofullerenes containing Cs, U, Ca, Ba, K, Na, Sr, and a number of actinide series and lanthanides were obtained.[36,37] These molecules possess unconventional electronic and structural properties.

It is interesting to discuss the properties of fullerenes with encapsulated radioactive atoms. At accidents in nuclear power stations (e.g., the Chernobyl accident) the possibility of endohedral metallo-fullerene formation may not be ruled out. The produced complexes represent an increased ecological danger since these radioactive particles possess extremely high penetration. More details about fullerenes can be found in Reference 38.

9.1.2 Modification of Polymers

Radiation modification of polymers proceeds as a result of a number of radiation chemical processes: cross-linking, a formation of transverse bonds between polymer chains; destruction, a dissociation of bonds in the main chain and side groups; radiation polymerization; and copolymerization. The regularities of these processes are described more fully in Reference 39.

9.1.2.1 Cross-Linking of Polymers

Irradiation of polymer materials gives rise to formation of spatially cross-linked polymer structures. In this case the molecular mass of polymer increases and an insoluble fraction, called a gel-fraction, appears, This process changes mechanical, physical, and chemical properties of polymers used for practical purposes.

Radiation modification of cable insulation and that of wires is widely used in industry.[40] A low heat resistance of thermoplastics and elastomers widely used as electro-insulation materials can be increased by radiation cross-linking of material. To produce radiation-modified insulation, polyethylene and polyvinylchloride are used most often. To a lesser extent, such treatment is applied to insulation of ethylene-propylene resin, chlorosulfonated polyethylene, butyl-rubber, etc. To reduce a dose rate relevant for cross-linking, which is especially important for polyvinylchloride having a low radiation-chemical yield of cross-linking, additions which sensibilize cross-linking are introduced.[41,42]

Radiationally modified materials are those resistant to thermal aging and can perform for a long time at temperatures of up to 150°C; they exceed the analogous characteristics of insulation from fluoropolymers and silicone. The advantages of radiation modified insulation materials also include reduction of wastes in processing raw materials, lower energy costs, the possibility of conducting continuous treatment extrusion, and improved rates of radiation processing. This results in reduced economic costs compared with chemical cross-linking techniques of insulating material.[43]

The change in the viscosity of a polyethylene melt under irradiation is used for polyethylene foam production. In order to produce a high-quality material, the foaming must take place at the optimal polyethylene viscosity. In the case of unirradiated polyethylene at temperatures over the melting temperature, the viscosity decreases sharply, while for irradiated polyethylene the viscosity decreases approximately according to the linear law.[44] Therefore, decomposition of the pore-forming agent (usually diazodicarbonamide is used) proceeds at temperatures when the viscosity of polyethylene is optimal.

The advantages of the radiation-chemical technique of producing foamed polyethylene as compared to the chemical method are associated with continuous industrial process, better material homogeneity, and economy of a cross-linking agent. High amortization, thermoinsulation, and sound-absorption properties; low water absorption; possible reprocessing by thermoformation and mechanical working; sufficiently high chemical and temperature resistance — these are the qualities of radiation-produced foam polymaterials that determine the areas of their application (packaging for products, construction, car building, etc.).[45]

"The memory effect", a phenomenon which is characteristic of irradiated polyethylene,[46] is used for producing heat-shrinkable tubes and films. The radiation cross-linked polyethylene possesses resinlike properties at temperatures higher than the melting point. This trait is manifested in reconstruction of the initial form at repeated heating of a polymer sample preliminarily heated up to the melting temperature, deformed, and then cooled in this state. To produce heat-shrinkable products, polyvinylchloride, silicon rubbers, polyisobutylene, propylene, polyvinylidenefluoride, copolymers of ethylene with ethylacrylate, and other polymers can be used as well. On the industrial scale, heat-shrinkable materials are utilized for packaging food stuffs, producing integral casings, connecting and sealing joints in different systems, and protecting electric circuits and pipings against the effects of aggressive media, etc.[47]

Radiation cross-linking of elastomers is usually a called "radiation vulcanization". The process is used to produce thermally stable auto-adhesive insulation tape and in the tire-manufacturing industry. Radiation vulcanization of tires does not bring about tangible advantages over the conventional thermal technique; however, preliminary irradiation of partially finished products of separate elements of tires gives them greater cohesive strength. This allows increased product shape stability with a decrease in mass.[48] Irradiation of moulding resin before assembling in order to increase its strength is widely used in tire-manufacturing processes.

To prepare a heat-resistant, self-adhesive for electrical tape insulation, use is made of radiation vulcanization of boron-containing polysiloxane rubber. The tapes produced possess good auto-adhesion, high water- and frost-resistance, and long-term operability at high (~250°C) temperatures.[49]

9.1.2.2 Destruction of High-Molecular Compounds

A process of radiation destruction of high-molecular compounds also finds wide application. In a number of cases, it is necessary to use polymers of a given molecular weight. At the same time, when polymerizing some monomers (e.g., polytetrafluoroethylene and polyethyleneoxide), it is difficult to obtain a polymer with a certain molecular weight. Exposing such a polymer to radiation, it is possible to regulate its molecular weight. This process has been realized on an industrial scale for polyethyleneoxide.[50] Polytetrafluoroethylene, with low molecular weight and used as a lubricant and material for coatings, is also obtained by use of this technique.[51]

Irradiation of a biomass containing cellulose (e.g., sawdust, waste paper, wood chips, corn-cob rods, clay) is a possible technique for increasing effectiveness in processing. For instance, under the action of high-energy radiation on cellulose, a decrease of the molecular weight of a polymer chain, and the effect of structure amorphousness, destruction of the lignin-cellulose complex and formation of monomer and oligomer hydrocarbons take place.[52] These changes allow an increase of the effectiveness of cellulose processing by hydrolysis in acidic medium or fermentative hydrolysis.

The γ-irradiation of wheat straw up to the dose rate of 2 MRad allows the glucose yield to be raised by more than three times and the total yield of hydrocarbons by over four times after acidic hydrolysis.[53] Similar results are obtained under fermentative hydrolysis of irradiated cellulose. It has been shown that in this case a noticeable increase in glucose yield also happens.[54] A degradation of cellulose-content materials improves their digestibility as fodder for animals.[55] A further chemical, microbiological, or fermentative processing of glucose allows a large amount of chemical substances produced currently from hydrocarbons to be obtained. The wide availability and large amount of inexpensive and reproducable cellulose-content wastes demonstrates the potential of their utilization in place of oil as a raw material for producing various chemical products.

Production of nuclear filters, thin polymer films with through pores of 10^{-3} to 10 μm in size, is based on the degradation of polymers under interaction with accelerated heavy ions or fission fragments in nuclear reactors.[56] Under irradiation with heavy particles, small-size domains are formed which have a changed chemical and physical structure and are called "latent tracks". They can be discovered by various means, in particular by selective etching of irradiated polymer. A formation of a hollow channel occurs owing to the difference in solving rate or as a result of the difference in the reactivity of irradiated and unirradiated parts of the polymer. Through holes in a polymer film can be obtained if a latent track extends over the whole material thickness, and the etching speed V_t is higher than the etching rate of the unirradiated material (V_m). The track geometry is determined by the ratio V_t/V_m which is often called selectivity or sensitivity. A track etching rate depends on the material, parameters of the particle which formed the track, conditions of storage, and etching. To a certain extent, a track geometry may be controlled using external effects such as ultraviolet radiation, temperature, etc.[57]

As materials for nuclear filters, routine use is made of polycarbonate or polyethyleneterephthalate films of ~5 to 10 μm thick. Besides, nuclear filters are obtained on the basis of polyamide, fluoroplaste, and polypropylene.[58,59] Dispersion of pore diameters of nuclear filters made of polyethyleneterephthalate with a diameter of 10^{-2} μm is equal to 15% and with a diameter of $5 \cdot 10^{-2}$ to 10 μm does not exceed 2%. Porosity is usually 10%.[57] High thermal and chemical resistance, biological passivity, and stability of the polymer properties used for nuclear filter production cause their wide practical application. Purification of air and process fluents for microelectronics, protection of respiration organs against aerosols' effect, cleaning of virus vaccines, fractioning, and blood examination are only some prospective areas of nuclear filters application.

9.1.2.3 Polymerization

A radiation polymerization possesses a number of advantages compared with conventional methods. These include absence of impurities in the form of catalysts or residues of the initiators of polymerization, high rate of initiation, independence of initiation rate on temperature, possibility of initiating polymerization in a solid state and unconventional conditions, and, lastly, the ease of process control. Some of these advantages are used in practice. Polymerization in the solid state is used to produce optical products (lenses, prisms, contact lenses, etc.).[60] Monomers of polymethylmethacrylate or other polymers are placed in a mold, frozen at a temperature close to the temperature of vitrification, and exposed to radiation. As a result, high-quality (absence of cracks and bubbles, high transparency) massive optical products are obtained.

Formation of high-molecular, strong-branched, and cross-linked polymers as a result of radiation polymerization of unsaturated oligomers (subpolymers having a molecular mass larger than that of monomers but less as compared with polymers, i.e., their chain length is small), and oligomer-monomer mixtures is called "polymerization" or "radiation curing".[61] The process of radiation curing is used widely in producing polymer coatings on various materials (metal, plastics, wood, paper, etc.). Finishing of wooden surfaces; applying decorative or protective coatings on construction parts of asbestos-cement or cement, metallic surfaces, and paper; curing of dyes in the printing industry; fabrication of magnetic coatings for videotapes and floppy-diskettes — these are the areas where the radiation curing of coatings has found industrial application. For more details on the application of this process, see Reference 63.

Radiation grafting copolymerization is used to modify polymers and inorganic materials with the aim to have their operational properties improved. As a result of such treatment, new and unique properties of materials are rendered. Formation of chemical bonds between a support (material to be grafted to) and substrate (modifying agent) allows purposeful development of materials combining the properties of the support and substrate. Other techniques usually fail to combine these properties.

The possibilities for modifying properties of different materials by this technique are rather extensive. Examples of such material modification include a change of hydrophilic and hydrophobic; an increase in adhesion to different substances; implantation of active centers into material possessing ion-exchange, complex-producing, or catalytic properties; an increase in mechanical, thermal, chemical, and radiation resistance; an increase in fire resistance, etc.

Grafting of acryl acid to polyethelene with the aim to improve adhesion to aluminium is utilized in production of packing material for food products.[64] Modification of textile fibers of natural and synthetic fibers by radiation grafting allows the addition of waterproof, refractory, and antistatic properties to them.[65-67]

Propylene with crossed-linked *p*-styryl-diphenylphosphine was used to prepare a metal-complex catalyst.[68] Grafting of acrylic acid, acrylamide, acrylonitrile, and others to various polymers helps to obtain ion-exchange membranes with high selectivity.[69-72] Radiation grafting is used widely for synthesis of chromatographic sorbents.[73]

The regularities of radiation-grafting copolymerization are discussed in References 74 to 77. Utilization of this process in radiation bioengineering and synthesis of sorbents for decontamination of radioactive wastes and purification of industrial wastes are described in respective sections of this chapter.

9.1.3 Development of Composite Materials and Encapsulation of Dry Substances

The processes considered in the previous subsection often proceed simultaneously. Owing to this fact, it is possible to develop new composite materials possessing a small mass; high stiffness; improved impact strength, size stability (small coefficient of thermal expansion), and strength, etc.

Radiation polymerization of monomers of wood leads to improving its operational properties: improved hardness and wear resistance, increased resistance against putrefication, and decreased water absorption.[78] As a monomer, usually methylmethacrylate is used. In this way soft and hard types of wood materials are modified, as well as the tropical timber of which sport fittings and other items are produced.[79,80]

Modification of concrete, asbestos-cement, tufa, paper, and other materials by radiation polymerization permits production of concrete-polymer[78] and paper-polymer[81] materials possessing higher resistance against external effects and enhanced properties. This technique is also used to protect pieces of fine art and archaeological findings.[82]

Radiation polymerization activation of dispersion fillers and pigments can change physical and chemical properties of polymer compositions. The causes of such influence may be a change in the character of crack formation and propagation, formation of the structural spatial frame (including a filler), and decrease of the hopping of the elasticity modulus at the filler-matrix interface.[83]

An increase was observed in the film strength of laquer materials using modified (e.g., by polystyrene, polymethylacrylate) fillers (chalk, TiO_2, MgO, etc.).[84,85] Similar results were obtained for polyvinyl chloride[86] and elastomers.[87] Adding a polymer coating on solid particles by radiation polymerization decreases the aggregation of particles of a micron dimension and increases the time of action (dissolution) of minerals.[88]

Radiation is also used to prepare and modify composite materials with a polymer matrix reinforced with carbon or glass fibers.[89] These materials possess a unique combination of such properties as high stiffness and low coefficient of thermal expansion. Radiation modification of ready-made composites leads to different changes in their mechanical behavior.[90-92] Irradiation of composites leads to a decrease breaking and flexural in strength. Modulus of transverse elasticity and breaking strength of fiberglass-reinforced plastics begin to decrease already at 20 MGy, while for fiber carbon-reinforced plastics changes are observed at dose rates 2.5 to 5 times higher.[93,94] Compression strength increases at low temperatures and decreases at high temperatures.

The effect of radiation on the strength of fiber carbon-reinforced plastics, fiberglass reinforced plastics, and organic plastics depends on both the surface properties of the resin-fiber interface (possibility for adhesive bonds to be formed and dissoluted) and resin properties. A change in the elastic properties of adhesive and substrate can give rise to creation or decrease of internal stresses in the contact area and to a change in the effectiveness of adhesion due to a small adjustment of interacting molecules. The adhesion strength is affected not only by the energy of the adhesion interaction but also by the mechanism and kinetics of activation or tunneling formation of stable centers of a new phase (broken adhesive bonds) of a subcritical dimension. Further growth of such a nucleus of the new phase proceeds in a nonactivation way. Also possible is a soliton (or dilaton) mechanism of bond dissociation.[95]

The changes in adhesion strength in composite materials (resistance to stratification) as a result of irradiation have been considered.[96-98] A sharp decrease in the adhesion strength of irradiated composite material (threshold character of its decrease) was explained by the influence of commensurability of adhesive and substrate structures on adhesion.[99] Even with an insignificant change of adhesive stiffness, the composite makes a transition by hopping to a state with fewer adhesive bonds and the energy of internal stresses increases sharply.[100,101]

Lately, considerable interest has arisen in carbon-carbon composite materials. These materials possess a low coefficient of thermal expansion and better strength characteristics than those of graphite with lower material density.[102]

9.2 RADIATION-CHEMICAL PROBLEMS OF THE NUCLEAR AND RADIATION-CHEMICAL INDUSTRY

Intensive development of nuclear power demands the solution of a number of problems associated with radiation resistance of applied materials, reprocessing of nuclear fuel, decontamination of radioactive wastes, etc. Solutions to these problems is based on understanding the regularities of radiation chemistry of solids.

When developing devices used in the nuclear and radiochemical industry, it is necessary to take into account that under the effects of irradiation a change in operational properties of the structure of materials occurs. In reprocessing nuclear fuel and radioactive wastes, it looks promising to use adsorption methods of purification and decontamination by ion exchange and complex-producing sorbents. To dispose of high-level wastes safely, use is made of their immobilization in different solid matrices, which will be considered in this section.

9.2.1 Radiation Strength of the Structure of Materials

Utilization of different materials operating in the field of ionizing radiation demands knowledge of their radiation resistance. Radiation resistance is characterized by the level of change in operational properties (strength, heat conductivity, electrical insulation properties, and others) referred to as the value of the absorbed radiation energy.

Materials such as metals, alloys, and ceramics possess high radiation resistance (noticeable changes of properties are observed at dose rates over 1000 MGy). A change of properties in these materials is associated mainly with a formation of different kinds of defects in the structure under interaction with radiation.[103] It can be, for example, radiation swelling (expansion of material's volume under irradiation) caused by a formation of vacancy pores. At temperatures close to half of the melting temperature of the material, annealing of radiation defects and the return of the material to the initial state take place. A change in mechanical properties is also determined by formation of defects. For example, an increase in the yield limit (load at which the specific residual deformation of material is 0.2%) is associated with fixing dislocations on the created defects.

Transition from the viscuous character of metal breaking to the brittle one under irradiation is associated with the processes of segregation of defects and impurities in the grain boundaries which may lead to material cracking along grain boundaries. Low-temperature and high-temperature radiation embrittlement of metals is observed. The high-temperature radiation embrittlement of metals is associated with aggregation in grain boundaries of gaseous products (e.g., helium) caused by nuclear reactions. The same reason seems to cause a radiation swelling of uranium (a change of the sample shape accompanied by a decrease in its density).

A change of properties in alloys under irradiation occurs due to the processes of ordering and disordering of their structure as a result of radiation-induced diffusion mobility of atoms. Phase transformations in alloys stimulated by irradiation lead to significant changes in their physical properties.

Moreover, the influence of radiation on the anticorrosion properties of a substance is possible. Considerable metal corrosion in water under irradiation can bring about destruction of structures and contamination of cooling agents with radionuclides generated by the failure of protective oxide films.[104-106] The growth of the corrosion rate is associated with a change of oxide-reduction electrochemical reactions owing to the product of water radiolysis and dissolved substances, as well as formation of nonequilibrium scavengers and defects in the oxide film.[107,108] The presence of oxygen, chlorine ions, pH-medium, and temperature also strongly affect the rate of radiation-induced corrosion.

Radiation resistance of polymer materials is determined mainly by the processes of cross-linking, destruction, and oxidation under irradiation. Noticeable radiation damages of polymer materials are observed at dose rates 10^4 to 10^6 Gy, which shows their low radiation resistance.[109] A change of mechanical properties of polymers characterized by fracture strength, relative extension at stretching, hardness, etc. depends on the effectiveness ratio of radiation cross-linking and destruction of material.[110,111] When the cross-linking of polymers prevails, the relative extension at rupture decreases with increasing dose rate, and elasticity modulus, hardness, breaking strength increase. It is characteristic of destructing polymers to exhibit a decrease in strength and durability under extension and for residual deformations to increase under compression. Such changes are characteristic of all types of ionizing radiation, although the quantitative parameters of dependences may differ substantially.[112,113]

Qualitative differences are observed when destruction of a polymer happens as a result of radiation-initiated oxidation.[114-116] In this case, if the reaction proceeds in the regime of internal diffusion (reaction rate is limited by diffusion of oxygen into the volume of material), then the heterogeneity of polymer failure is observed and, correspondingly, a formation of composite material (two-layer structure with an external oxidized polymer layer and cross-linked internal volume). The dependence of the mechanical properties of such materials on the rate and value of the absorbed dose has a complicated character related to increasing levels of polymer film failure, dissimilarity with the rate of growth, and value of the absorbed dose.

The influence of radiation on electrical properties (electrical conductivity, breakdown voltage, angle tangent of dielectric losses) of polymers widely used as insulating and protective materials for cables and wires in equipment operating in an ionizing radiation field is related to formation of the products of their radiolysis.[117] Estimation and prediction of the operability time of electrical insulation materials under radiation effect have been determined.[118,119] It should be noted that the limiting dose rates at which electro-insulation materials retain their electrical properties are, as a rule, higher than the dose rates at which their mechanical properties degrade significantly. A great amount of data on radiation resistance of various materials is given in References 120 and 121.

9.2.2 Reprocessing of Nuclear Fuel and Radioactive Wastes

Selective separation and concentration of radioactive elements are the basic aims of the nuclear fuel cycle. As is known, the nuclear fuel cycle consists of a number of processes, including extraction and concentration of uranium from ore; reprocessing of irradiated fuel to separate uranium, plutonium, and other radioactive components of the irradiated fuel; utilization of nuclear wastes; purification of reactor feed-water and the reactor and storage pond; cooling water from radioactive corrosion products; and minimizing consequences of accidents releasing radioactive elements to the environment.

9.2.2.1 Solvent Extraction

Currently, the principal method of extracting and concentrating radioactive components in the nuclear fuel cycle is solvent extraction. In the capacity of extragents, different complex-generating substances are utilized, such as amines and phosphororganic compounds.[122] It is also possible to use macrocyclic compounds in the capacity of extragents.[123,124] Extragents must have high-effectiveness, complexing-ability selectivity, radiation resistance, stability of properties under conditions of irradiation of extraction systems, as well as easy regeneration in different acidic solutions.

Apart from the evident advantages of using solvent extraction, it has some disadvantages, such as utilization of hydrocarbon radiation-irresistant diluents, comparatively small values of distribution coefficient (ratio of the equilibrium-extracted concentrations element in the extragent phase and the solution reprocessed) which increases the number of extraction cycles and makes it difficult to reprocess solutions with low activity, an increase in the volume of liquid radioactive wastes, etc.

9.2.2.2 Sorption in Ion-Exchange Materials

A sorption technique is most suitable for utilizing radioactive wastes. The basics of this technique are sorption of radioactive elements by a solid polymer and inorganic sorbents possessing ion-exchange or complexing-ability properties. It allows considerable simplification and improvement of the effectiveness of selective extraction and concentration in reprocessing nuclear fuel and radioactive wastes. With the use of sorption technology it becomes possible to avoid the use of organic solvents. High distribution coefficients, the possibility of working in aqueous solutions of different acidities in the dynamic regime, ease of sorbent regeneration, high radiation resistance (especially of inorganic sorbents) — these are the properties which lead to the use of sorbents at different stages of the nuclear fuel cycle.

Reference 125 discusses the adsorption properties of inorganic ion-exchange materials (natural and artificial zeolites, oxides and hydroxides, metallic salts with variable valences). Such sorbents are used in a wide range of solutions with pH = 12 to 4 *M* of nitric acid content. The maximum values of distribution coefficients of actinides (Pu, Am, Np) for inorganic sorbents are observed at pH = 1–3. When extracting by means of hydroxide titanium, the distribution coefficients are equal to $4.198 \cdot 10^3$ and $9 \cdot 10^5$ for Am and Np, respectively. And, when extracting Pu by means of phosphate titanium, the distribution coefficient equals $5.62 \cdot 10^4$. A high level of extraction (~100%) of radioactive Cr, Sr, Ce from solutions has been achieved as a result of applying polyatomic antimony acid. To extract Ru, the best result has been achieved by applying magnesium dioxide. The γ-irradiation of sorbents with a dose rate of 10 MGy did not lead to sufficient changes in sorption properties of these sorbents. Low selectivity and capacity are disadvantages which constrain the area of their application.

Polymer ion-exchange materials possess a large sorption capacity (up to 10 μeq/g, (microequivalents per gram.)) as compared to inorganic ion-exchange materials and are used to purify liquid radiation wastes.[126] The ion-exchange resins contain acidic groups of sulphoric and other acids grafted to a polymer matrix. For instance, the ion-exchange resin, containing sulphate groups in a cross-linked polystyrene matrix, has a capacity of 4.8 μeq/g and completely removes Co^{60} when passing through $6 \cdot 10^4$ column volumes of aqueous radioactive wastes. The distribution coefficient in this case is $3 \cdot 10^3$ under extraction of Co^{60} from aqueous solutions containing 2 g/l of magnesium salts at lower concentrations. Sorbents containing amino groups effectively extract uranium from sea water.[127] The disadvantages of ion-exchange materials are the slow kinetics of ion sorption, contamination of the purified solution by ion-exchange products (i.e., they bind only a cation of metal which leads, for example, to an increase in the acidity of solution), low selectively, and little effectiveness in solutions with high acidity.

9.2.2.3 Complex-Generating Sorbents

Macrocyclic polyethers or crown-ethers are capable of effectively generating complexes with cations of different metals. The stability of the complexes generated is substantially greater than that of the compounds where ligands are linear analogues of crown-ethers with the same number of donor atoms. This phenomenon was called a "macrocyclic effect".[128] Selectivity of the complexing-ability of crown-ethers to a large extent is determined by parameters of their structure, sizes of a polyether ring, nature of substitutes in a macrocycle, type of donor atoms, stiffness of molecule, etc. By changing these parameters, it is possible to obtain macrocycles with formulated properties. Properties such as high selectively of formation and stability of complexes with cations of metals make crown-ethers possible bases for developing sorbents.

Complexing properties of immobilized crown-ethers have been described.[129,130] Immobilization of crown-ethers on the polymer or inorganic support sufficiently changes the effectiveness and selectivity of their complexing-ability. This happens as a result of a change in the conformation and stiffness of the macrocycle, specific features of solvatation processes owing to surface presence, cooperative effects (joint action of two or more neighbouring crown-ether groups, etc.). For instance, by using 18-crown-6 grafted to silica gel, it is possible to selectively extract Ba and Sr from aqueous solutions containing sufficiently larger amounts of Ca and Mg. Nitrogen-content crown-ethers effectively separate transition metals (Zn, Co, Ni, Cu).

The cross-linking of polymer with grafted benz-18-crown-6 when the crown-ether stays in the form of a complex increases the selectivity of Cs extraction in the presence of K (cooperative effect). An increase in the extraction level of Sr from aqueous solutions containing nitric acid grafted to silica gel DCH18K6 with increasing solution acidity (synergism of complexing ability in acidic media) was observed.

Specific features such as high effectiveness and sorption selectivity of metallic salts, the possibility of being used in acidic solution, the large amount of sorption-regeneration cycles, the extraction of metals in a salt form (i.e., together with anion, in contrast to ion-exchangers) are the major merits of sorbents with immobilized crown-ethers. A relative low capacity (0.5 to 5 μeq/g) and complex synthesis which lead to high costs are considered disadvantages of such sorbents. Use of kryptands and calixarenes, macrocyclic compounds possessing still higher effectiveness, and selectivity of complex generation with ions of metals will allow improvement of the quality of the sorbents.

9.2.3 Disposal and Decontamination of Radioactive Wastes

Improvement of disposal and decontamination techniques for radioactive wastes is not only a scientific and technical problem but also an important social and political issue. Risk assessment for human health and undesirable consequences for the biosphere and geosphere determine the level of radiation safety in disposal of radioactive wastes.[131]

Radioactive wastes contain radionuclides with different half-life times. The major fractions of activity are Cs^{137} and Sr^{90} whose half-lives are 30.7 and 29 years, respectively. Owing to a large value of half-lives, long-term isolation of wastes is necessary to reduce their activity up to a natural level (natural background). At present, the principal method of radioactive waste disposal is to immobilize the waste in solid matrices, since the storage of liquid radioactive wastes does not guarantee safety. The capability of different substances to retain radionuclides safely and effectively is shown in Reference 132.

To solidify wastes of intermediate level activity, a technique of bituminization is applied. Under waste bituminization, mixtures are obtained containing 50 to 60% by weight salts with a low rate of radionuclide leaching. To immobilize wastes, vitrification is also used, as well as in the production of ceramics, incorporation of solidified wastes in metallic matrices, concretization, etc.

Application of boron-silica glass for immobilizing high-level radioactive wastes is based on its thermal and radiation resistance, chemical duration, capability of incorporating different radionuclides, as well as mechanical strength of vitrified wastes.[133] Glasses utilized in the U.S., France, and Germany contain 45 to 47% by weight SiO_2, 5 to 20% by weight B_2O_3, 1 to 20% by weight Na_2O, and some proportion of other oxides. Glass can effectively retain over 40 different elements in a wide range of concentrations contained in the wastes. The vitrified wastes are up to 30% by weight.

Phase separation and crystallization of vitrified wastes under heating due to radionuclide decay does not lead to a decrease in their chemical durability. Irradiation of glass with alpha- and beta-particles, gamma-rays, and neutrons during radionuclide decay does not change its mechanical properties. The chemical durability of glass is determined by corrosion processes. The formation of a surface layer decreases glass dissolution which improves its chemical durability. These questions are treated in more detail in Reference 134.

The immobilized high-level wastes, in spite of their small dose rate and large dimensions as compared to standard radiation sources (for example Co^{60}), may be used as a source of radiation,[135] with both utilization of nuclear wastes and their secondary application being achieved.

9.3 RADIATION PROCESSING IN MICROELECTRONICS

Wide use of electronic products in nuclear power engineering and cosmic technology requires an investigation of radiation influence on materials and products applied in electronics engineering. For example, the absorbed annual dose of ionizing radiation (protons, electrons, bremstrahlung radiation) behind the shielding layer 1 g/cm^2 is great enough both in the inner radiation belt of the Earth (300 to 10000 km above the surface) and the outer radiation belt (10000 to 60000 km) and come to about 10^3 to 10^4 Gy respectively.[136] The electronic equipment of nuclear power plants, depending on the location and protection, absorbs a dose rate of 10^4 to 10^8 Gy during 30 years of operation.[137] The resistance of electronics to the pulsed radiation effect is an important factor in military technology. At the same time, the semiconducting devices, as a rule, have low radiation resistance. Owing to this fact, determination of the mechanism of radiation influence on semiconducting devices and materials is a necessity for increasing their radiation resistance. Radiation methods are also used in producing and modifying semiconducting instruments.

9.3.1 Radiation Resistance of Elements and Devices

Radiation damage to semiconducting devices is determined by formation of defects of a different type under radiation interaction with semiconducting materials.[138,139] Defects producing volume and surface levels in the forbidden band are the centers of trap, and recombination centers (see Chapter 1) determine the basic electrical and physical characteristics of semiconducting materials. They influence the concentration of charge carriers, mobility, lifetime of charge carriers and correspondingly the diffusion path length.

9.3.1.1 Semiconducting Diodes

Let us consider the principal regularities of radiation influence on the properties of semiconducting devices.[140,141] Degradation of semiconducting photoconvertors (solar cells) based on *p-n* transitions in silicon during interaction with radiation is characterized by reduction of short-circuit current. As a result the maximum useful capacity of the photoconvertor decreases, too. The main reason for such changes is the dependence of the diffusion path length of minority current carriers on their lifetime. By the same reason, the spectral distribution of the photosensitivity of the element changes. Maximum sensitivity shifts toward the short-wave band of the spectrum.

A low radiation resistance of *p-n* transitions in silicon (changes are observed at relatively low dose rates of 10^3 to 10^4 Gy) requires the use of different routes to protect them. To add a protective layer of glass or quartz is considered a passive way. An active method is associated with silicon alloying by means of various chemical elements, increasing the radiation resistance of *p-n* transitions.

The increase of backward and decrease of direct conductivity of *p-n* transitions (degradation of rectifying properties) during high-energy radiation are characteristic of practically all types of diodes (rectifying, high-frequency, etc.). The volt-ampere characteristic of diodes is determined by concentration and lifetime of nonmajor carriers generated thermally. A change of these parameters on irradiation is a cause of diode degradation. In addition, the rectifying properties of diodes deteriorate upon sample irradiation because of the lifetime reduction of minority carriers, the change of the level, and sometimes the character of alloying.

Irradiation also allows the control of limiting the working frequency of a diode. This happens due to the dependence of diode frequency characteristic on the lifetime of minority carriers. The radiation method of controlling the lifetime of carriers is utilized in production of high-frequency semiconducting devices. Examples will be given in the following section.

Tunneling diodes are characterized by high radiation resistance. This is associated with the fact that the *p-n* transition, created in a strongly alloyed semiconductor (the concentration of electrons and holes $\sim 10^{18}$ to 10^{19} cm^{-1}), is the active element of a tunneling diode. It means that in such a sample there is already a considerable amount of impurities and, consequently, additional radiation effects do not bring about significant changes. Since the processes in the tunneling diode are caused by the majority carriers, the theoretical limit up to which the tunneling diodes can work is determined by the Maxwell relaxation time. This relaxation time is inversely proportional to conductivity. Under interaction with radiation, a decrease is observed in the ratio of the maximum and minimum values of the incident part of the volt-ampere characteristic. At the same time, a decrease of the maximum value of the negative differential conductivity of tunneling diodes is observed. These parameters are determined by spreading of the *p-n* transition as a result of the creation of defects. A change of the frequency is characterized by a change in the mobility of carriers. Such changes in semiconductors become noticeable only at rather large absorbed radiation dose rates.

9.3.1.2 Transistors

Transistors are semiconducting instruments with two or more *p-n* transitions intended for amplifying and generating electrical signals. Bipolar transistors change under irradiation of the majority of their parameters. The major parameter is the static coefficient of current transfer (current gain) according to a scheme with a common emitter. The decrease of this basic parameter is associated mainly with decreasing lifetime of the minority current carriers. This occurs as a result of increasing recombination processes in the radiation-formed recombination centers.

In transistors produced by use of planar processing, the lifetime of the carriers is mainly determined by surface states. Irradiation in this case creates extra fast surface states as recombination centers. In addition, both fast and slow surface states are formed and are responsible for trapping current carriers. An increase in the number of recombination centers leads to an increased rate of growth of surface recombinations. The radiation influence on the number of trapped centers leads to a change of the surface state charge and the value of the surface potential. The consequence of this fact can be both change of the surface recombination rate and the formation of inversion near-surface layers of the type of conductivity which also strongly affects the working parameters of transistors.

The change of surface states especially affects the characteristics of unipolar transistors. Let us consider, for example, the peculiarities of such an influence on MOS (metal oxide structures).[142,143] The electrical current in such a transistor flows from the electrode, called the "source", to the electrode, called the "drain", through the narrow region of the semiconductor (the "channel"). Above the channel, the control electrode, the "control gate", is located and can be used to alter channel conductivity. Therefore, a change of parameters of such structures is determined by processes occurring at the semiconductor — insulating dielectric interface (e.g., Si-SiO_2). Formation of a positive spatial charge in the dielectric layer as a result of irradiation, simulating the channel conductivity, is associated with a shear in the threshold voltage (a voltage at which channel "overlapping" happens) toward the remaining large values. The field formation of a positive volume charge takes place due to the capture of the radiation-produced holes over the surface states (biographic or radiation-produced). The increase of the surface recombination rate because of the density growth of the surface states on irradiation is associated with decreasing steepness of volt-ampere and volt-capacity characteristics of such transistors.

9.3.1.3 Integral Circuits

The radiation resistance of integral circuits will be determined by the radiation resistance of its constituent discrete elements. The main factors which bring about a change in integral circuit parameters under radiation may include:

1. Change of the concentration and mobility of the main current carriers in semiconducting materials with a low concentration of the alloying admixture
2. Decrease of the insulating properties of dielectric layers
3. Decrease of the lifetime of charge carriers as a result of formation of volume and surface recombination centers
4. Formation of a volume electrical charge in the dielectric-semiconductor interface

Calculation and determination of the radiation resistance of integral microcircuits is rather complicated. These issues are described in more detail in References 144 and 145.

9.3.2 Radiation Methods Producing Semiconducting Instruments

Radiation methods are used to modify and produce semiconducting instruments. First of all, we will discuss the process of lithography. Besides radiation, alloying of semiconductors allows the production of semiconducting instruments with relevant characteristics (sometimes exceeding the parameters of instruments manufactured by use of other methods).

9.3.2.1 Modification of Semiconducting Instruments

Radiation techniques for modifying semiconducting instruments are based on the processes of their radiation damage, as described above. In several cases, the radiation effect on semiconducting instruments leads to improvement of their characteristics. A decrease in the lifetime of carriers upon irradiation with different types of radiation (γ-radiation, electrons, neutrons, protons, etc.) is utilized for decreasing recovery time and switching time of silicon diodes and transistors.[146,147] The radiation technique permits the control of the switching time of a semiconducting instrument once it is assembled.[148,149] The optimal dose rates for modifying transistors with γ-irradiation, fast neutrons,[148] electrons,[150] protons, and alpha-particles[151] are rather large. Irradiation by electrons is also used to reduce the switching time and losses in powerful high-voltage bipolar transistors. These transistors are used in fabricating displays (video display terminals),[152] to produce silicon-controlled rectifiers and fast recovery diodes,[153] and to increase the switching rate of the inner diode in powerful high-voltage MOS equipment.[154]

Radiation defects form donor and acceptors levels in the energy gap. By altering the concentration of defects it is possible with high precision and good reproducability to develop a material with relevant resistance and type of conductivity. Such a technique of alloying, despite the fact that annealing of radiation defects occurs at temperatures of 250 to 300°C, is more suitable and inexpensive than ordinary alloying with chemical elements.[155] The technique is used on a large scale to produce semiconducting instruments.

9.3.2.2 Lithography

The process of lithography is a part of the production processing of semiconducting instruments and integral microcircuits. It is based on the capability of high-molecular compounds, called resists, to alter their properties under action of different types of radiation (see Section 9.2.2). As a radiation source, ultraviolet and X-ray radiations, electrons, and heavy charged particle beams are used.

To create high-precision reliefs of future circuits in film layers or directly on a support, a contact or projective lithograph is widely used.[156,157] Silicon covered with an isolating layer of silicon dioxide is coated by resist (0.1 to 1 μm thick) and is exposed to radiation through a mask-mold with the plotted pattern of an opaque-to-radiation material. Once the resist is developed, a complex pattern of the circuit with very small element dimensions is formed on the surface of the semiconducting crystal. In the case of a cross-linking polymer, as the resist is developed (removal of individual layers of the resist by means of dissolving in chemical agents or by plasma treatment), the nonirradiated sections are removed. Such a resist is called negative. If the polymer destructs, then during development the irradiated sections of the polymer are removed and the resist is called positive.

After the image is formed, etching of the isolating layer is undertaken (usually by hydrofluoric acid) and the following operations of producing semiconducting instruments are carried out: oxidation, diffusion, coating metallic films, etc. The smaller the size of the elements, the denser packing and the shorter the switching times can be achieved.

A resist is characterized by sensitivity, resolution, and visibility. A resist's sensitivity to radiation is determined by the dependence of its thickness after development on the exposure dose rate. When irradiating with an electron beam, the exposure dose rate is expressed in coulombs per square centimeter (C/c^2), and with X-rays, in joules per square centimeter (J/C^2). For a positive resist the measure of radiation sensitivity is the minimum dose of exposure necessary for complete removal of the irradiated sections of the resist. For a negative resist, such a measure is the dose at which the thickness of the irradiated sections of the resist is 50% of the initial.

The magnitude of radiation sensitivity depends on the radiation-chemical yields of cross-links and to destruction of the polymer used as the resist, the initial mean-valve molecular mass, total mass stopping power of the substance, character of solvent, time and temperature of development, and resist thickness.[158]

The recalculation of exposure to absorbed dose, which determines the radiation-chemical transformations of a substance, is conducted using the regularities of radiation-substance interaction (see Chapter 2). Under electron irradiation, the absorbed dose is associated with energy losses of electrons per unit path length and the X-ray absorptions to the mass energy-absorption coefficient. The heterogeneous character of the considered system, as a rule, is not taken into account.

For the purpose of lithography, electron beams with energy from 5 to 30 keV and X-ray radiation in the range of wave lengths from 0.1 to 10 nm are usually used. The radiation sensitivity of resists varies over a wide range.[156,158-160] For example, the sensitivity of positive resists on electron radiation is equal to $4 \cdot 10^{-3}$ C/cm^2 for polymethylmetacrylate and $2.5 \cdot 10^{-8}$ C/cm^2 for polymethylcrylate anhydride. When using X-ray radiation, the sensitivity of the positive resists of polymethylmetacrylate is equal to 1.0 to 2.5 J/cm^2 and its derivatives, $5 \cdot 10^{-2}$ J/cm^2. For negative electron resistive materials, sensitivity for polyvinylbenzylchloride is $5 \cdot 10^{-7}$ Cl/cm^2 and 10^{-8} C/cm^2 for polybutadiene containing epoxy groups. Under X-ray radiation, the negative resist (polybutadiene containing epoxy groups) has a sensitivity of $5 \cdot 10^{-3}$ J/cm^2.

To increase resist sensitivity, polymer materials with a large radiation-chemical yield of cross-links or destruction are selected. Also, heavy elements are incorporated into the material. Other important characteristics of resists are their visibility and resolution. Resist visibility is the tenth logarithm of the ratio of exposure dose rates which are obtained by extrapolating the linear section of the dependence of the relative thickness of the resist layer after development on the dose rate to the values of the relative thickness of the developed layer corresponding to zero and complete removal of the layer. Negative resists possessing higher sensitivity as compared to positive resists are characterized by lower visibility (0.5 to 2) than the positive ones (2.5 to 3). The low contrast of resists requires the use of high-contrast molds which constrains their application when the element sizes lie in a submicro region.

Resolution is the smallest size of the element which can be reproduced in a layer of the resist. Resolution capability depends on many factors but, in particular, on the type of radiation. A comparison of resolution capability vs. the type of source was carried out.[161] The use of visible region light allows a resolution capability of 1 μm to be achieved and resolution is limited by light diffraction.[162] By use of X-ray radiation and electron beams, a resolution of 0.1 μm is achieved. Resolution is constrained by diffraction and dissipation, respectively. Use of high-energy heavy-particle beams has some advantages compared to the traditional processing.[163] Besides a high resolution (0.01 μm), they assist in producing three-dimensional structures which greatly enhances the production of semiconducting instruments and microcircuits. To solve the problem of the optimal choice of a polymer used in the capacity of a resist (it must possess a number of completely different properties), a large amount of different polymers have been synthesized and investigated.[164,165]

For instance, it was shown that poly-1-methyl-styrene, deposited on the surface of silica-dioxide under exposure to γ-radiation or electron beam in a vacuum, is subjected to depolymerization up to a monomer. The process proceeds by a chain reaction.[165] The radiation-chemical yield of the monomer is about 100 molecules per 100 eV at room temperature. The conversion of the polymer may be significant. The use of such polymers as resists will allow the migration process to be controlled since it will be possible to exclude the labour-intensive procedure of etching a resist. It is obvious that such a technique of lithography is very promising.

9.4 BIOLOGICAL AND MEDICAL ASPECTS OF RADIATION PROCESSING

The study of the interaction regularities of ionizing radiation with biological objects is an important and complicated fundamental task demanding that interdisciplinary barriers be overcome. For example, small radiation-chemical changes in biopolymers (deoxyribonucleic acid, chromatines, proteins, lipids, etc.) lead to the majority of biological effects (reproductive death of cells, i.e., the loss of ability to reproduce, aberration of chromosomes, bringing about mutation and cancerogenesis; deactivation of enzymes, etc.). Biological objects are complicated organized systems and it is difficult, as a rule, to determine the mechanism of radiation effect on such objects. It may be expected that the application of nonlinear dynamic methods to this problem will be perspective.

This section will treat only some problems related to the field of radiation biology and radiation medicine. First of all they cover radiation sterilization of medical products and radiation treatment of food products. The effect of small radiation dose rates which cannot be predicted by extrapolating effects obtained at high dose rates is an important issue, particularly in light of the Chernobyl accident. Traditional radiochemical methods in bioengineering also will be considered.

9.4.1 Radiation Damages in Biopolymers

Radiation changes in biopolymers in real biological objects take place as a result of both the direct action of radiation (direct radiation absorption by biopolymer) and the indirect action of radiation (interaction of biopolymer with active products of water radiolysis, for example, hydroxyl radicals, hydrogen atoms, hydrated electrons, etc.). The structures of the most important biopolymers, nucleic acids and proteins, are well known and have allowed the primary directions of radiation-chemical transformations of these compounds to be established accurately.[167,168]

For instance, DNA is a polymer with a very high molecular mass consisting of nucleotides. A nucleotide, in turn, contains a purine or pyrimidine base connected with a 5-term hydrocarbon (deoxyribose) and a phosphate group. A pair of spirals, the main chain of which is repeated deoxiribose and phosphate chains, is connected by hydrogen bonds between bases. In a live cell, DNA is usually present in such a form.

The major types of damage in a DNA molecule are disssociation of the sugar-phosphate bond (fracture of strands). A dissociation of one or two strands may occur. The radiation yield of one-strand dissociation is 0.2 to 0.8; for two-strand dissociation, 0.03 to 0.09. Also, observation is made of various radiation damages of the bases, their stratification from DNA, dissociation of hydrogen bonds between circles, formation of inter- and intramolecular cross-links, disintegration of deoxyribose. The damage to DNC is the reason for the reproductive death of cells.

The structural base of proteins is a polypeptide chain consisting of sequentially connected amino-acidic residues. The peptide bond –CO–NH– is formed by a –COOH group of one and a $-NH_2-$ group of another amino acid. Proteins can consist of one or several such chains, connected by noncovalence bonds (for example, a hydrogen bond) or covalence disulfide bonds.

Under irradiation of proteins, degradation of the constituent amino acids is observed. Mainly, amino acids containing serum and aromatic rings or heterocycles degrade. Considerable conformational changes in proteins take place as a result of cross-links (aggregation), breakdowns of peptide bonds, and breakdowns of disulfide and hydrogen bonds. Such radiation-chemical transformations in proteins lead to a change of some biological properties of these molecules: antigen immunological specificity and catalytic selectivity of protein-pherments. A change of biological properties is possible only at rather high dose rates, some tens of MGy. It should be noted that to explain electronic properties of proteins and nucleic acids the methods of solid-state physics are used successfully.[169,170]

Under irradiation of chromatin (a complex compound of DNA with proteins), the radiation-chemical yield of one- and two-strand breakdowns in the DNA is more than an order less as compared to free DNA. Damages to the bases decrease, too. On the other hand, cross-linking of DNA and protein takes place.

Lipids (fatty acids, grease, steroids, phospholipids) are the important components of biological membranes. The action of radiation on lipids is closely connected with alteration of some functions of the biomembrane (for example, ion transport). The products of lipid radiolysis possess toxic properties which should be taken into account in the radiation treatment of food products. Lipids are characterized by the level of unsaturation, i.e., the presence of double bonds which are very sensitive to radiation. In the presence of oxygen, lipid hydroperoxides disintegrate into different aldehydes and ketones.

9.4.2 Sterilization of Medicinal Products

The sterilizing effect of ionizing radiation is based on suppressing the reproductivity of microorganisms, i.e., reproductive death. Death of microorganisms is of a probability character; therefore, the radio-resistiveness of microorganisms is determined from the function of the survived cells fraction vs. the radiation dose rate. This dependence, called the "survival curve", is described by the kinetics of the first order from the radiation dose.[171] Also, survival curves are observed with the so-called "shoulder of delay"; i.e., the exponential decrease of the fraction of the survived microorganisms starts not from the zero but from some final radiation dose.[172] This is explained by the work of reparation systems of the microorganism to remove radiation damages.

Effectiveness of the sterilization process is characterized by the dose at which the fraction of the surviving microorganisms is reduced by ten times. The dose of the decimal reduction (D_{10}) changes over a wide range (10^{-2} to 6 kGy) depending on the kind of pathogenic microorganism and irradiation conditions (medium, temperature, etc.).[172] A product is considered sterile when the fraction of the after-irradiation surviving microorganisms does not exceed 10^{-6} (sterility accuracy level).[173]

The standard of the sterilizing dose was established to be 25 kGy. The value was chosen using spores of Bacillus Pumilus AT CC 27142 (E601), the mostly radiation-resistant microorganism, as indicators of radiation sterilization.[171] The sterilizing dose is added from the dose relevant for sterilization of this standard microorganism up to a sterility accuracy level (18 kGy) and 40% of the margin for confident annihilation of possibly present radio-resistent microorganisms. Sometimes, doses of 50 kGy are used.[174]

Lately, the value of the sterilization dose has been subject to doubt. This is due to several reasons. The decrease of sensitivity of microorganisms at a low concentration of a microflores requires in the opinion of the authors[175] a revision of the procedure for determining the value D_{10} and the standard sterilization dose. Moreover, in real conditions there are various kinds of microorganisms present simultaneously, therefore, their resistivity has a nonmonotonous distribution which must be taken into account in determining a sterilization dose.[176] It was discovered that the *Acinetobacter* sp. has a higher resistivity and the dose 25 kGy is insufficient for confident sterilization.[177]

Under irradiation of products used in medicine, with a sterilizing dose rate it is necessary to take into account their radiation resistance. High radiation resistance of metals and glasses allows their multiple sterilization to be carried out. Sterilization of medicinal products of polymer materials, medicine preparations, and, particularly, biological tissues requires accounting for the radiation-chemical changes under irradiation. We note that the radiation resistance of a number of polymers such as polyethylene, polycarbonate, polystyene, polyvinylchloride, silicon rubbers, epoxide resins, etc. is rather high. This allows their multiple irradiation to be carried out.

Other polymers, for example polymethylmetacrylate, polypropylene, cellulose, and its derivatives can be irradiated only one or two times without changing the physical and chemical properties. In polytetrafluoroethylene, radiation-chemical changes are observed at dose rates less than the sterilization dose, which prevents their application.[121,178,179] A rather high radiation resistance of polymers allows wide application of radiation sterilization of various one-time medical products (syringes, catheters, systems of blood transfusion, wound dressing, and cross-linking materials, etc.).

Radiation sterilization is also applied to medicinal preparations. Solid drugs such as pills, powders, lubricants, pastes, creams, etc. have a rather high radiation resistance so that their radiation sterilization is carried out on an industrial scale.[180] Less radiation-resistant are medicinal preparations presenting aqueous solutions. It is suggested that a low-temperature sterilization be used for them which permits the radiation decomposition of such preparations to be reduced sufficiently.[181,182]

Radiation sterilization of biological tissues and preparations (bones, cartilages, vessels, valves, organs, blood, etc.), for which traditional thermal or chemical methods of sterilization are unsuitable, becomes difficult because of their high radiation sensitivity to a change of physico-chemical and biological properties. In several cases, the combined action of ionizing radiation and heat allows a significant reduction of sterilization dose.[183]

Besides, it is possible to use a sterilization process of controlled depth.[184,185] Use of low-energy electrons (0.25 MeV) allows the sterilization of medicinal product surfaces without changing volume properties. Under radiation sterilization of medicinal products, besides their microbiological safety, the question of toxicological and cancerogenic safety of irradiated medicinal products remains important.[186]

9.4.3 Radiation Treatment of Food Products

Radiation treatment is widely used to sterilize, pasteurize, and eliminate insects from food products, sprout inhibition of bulbs and tubes delay maturation of fresh fruits and vegetables. Extension of shelf life of some foods and improvement of the technology and quality of food products are improved too. Duration of storage of food products is extended and annihilation of pathogenic bacteria and organisms hazardous to health is achieved.[187,188] Sprout inhibition of onions, potatoes, garlic, carrots, and beets takes place at dose rates of less than 0.2 KGy. This effect is associated with direct damage to DNA and guinoid inhibitors of cell division created as a result of radiation.[189] Sprout inhibition of beets for several months (i.e., practically the whole season of storage) creates a sharp reduction of product losses.

To eliminate insects in grain, fresh and dried fruits, dehydrated fish and meat, dose rates up to 0.5 kGy are used. High radiosensitivity during critical steps of an insect's growth allows complete suppression of development and reproduction of insects under irradiation of contaminated material. Trichinella and tapeworm, parasites in pork, also are eliminated at this dose rate. A slow-down of physiological processes in fresh fruits and vegetables upon irradiation of up to 1 kGy delays their ripening by 5 to 10 days when stored at temperatures up to 10°C. In such a way, bananas, mangos, tomatoes, and other produce are treated to increase their lifetimes. Elimination of sporous bacteria, mustiness, yeast, and pathogenic microorganisms is achieved by irradiating food with a dose of up to 10 kGy. This allows the storage time of food to be extended by up to a month and food poisoning to be prevented. Moreover, the technological properties of food raw materials are improved. Juice yield from irradiated fruits increases, the cooking time of some dried vegetables decreases, the taste qualities of corn, beans, etc. are improved. For radiation sterilization of food and decontamination of food additives and ingredients, a dose of up to 50 kGy is used: under radiation treatment of food with dose 0.4-10 KGy is called radicidation, 0.1 to 8 kGy and radurization; 10 to 50 kGy radappertization.[190,191]

The International Atomic Energy Agency (IAEA), Food and Agriculture Organization of the United Nation (FAO), and World Health Organization (WHO) recommend the use of gamma-radiation with energy of up to 5 MeV (induced radioactivity does not appear) with an average level of absorbed dose of up to 10 kGy for radiation treatment of food.[192] Special attention has been devoted to demonstrating the harmless nature of irradiated products.[193] In this connection, chemical changes in food components have been studied intensively, and methods of investigating food products irradiation have been developed. Commercial use of the radiation treatment of food products in different countries is described in Reference 194.

9.4.4 Harmful Effects of Small Radiation Doses

Estimation of unfavorable effects under irradiation of biological objects with a small dose (several Grays) is an important problem. As far as the practical aspects are concerned, this importance is related to increased background radiation and its effect on human health in connection with nuclear bomb testing, disposal of radioactive waste, release of radioactive waste as a result of accidents at nuclear facilities, etc. In several cases the ability to extrapolate the magnitude of the effect from large doses to small ones demands the study of the effectiveness of small radiation doses and the determination of relative biological effectiveness or dose-equivalents. This is necessary to provide adequate radiation protection for those living in the presence of ionizing radiation sources, and their property.

The International Committee on Radiation Protection classifies radiation effects of small doses as stochastic and deterministic.[195] Stochastic effects develop at small doses, the probability of their creation increases with dose, and the severity of hazards does not depend on dose. Deterministic effects are characterized by threshold doses, and the probability and severity of hazards increase sharply once they are reached.

The high effectiveness of a small radiation dose was demonstrated when studying radiation effect on artificial lipid membranes.[196] It was shown that the same level of membrane damage occurs both under irradiation with a small dose (dose rate 10^{-5} Gy/min, dose $7 \cdot 10^{-3}$ Gy) and under irradiation with a dose 5000 times larger (dose rate 26 Gy/min, dose 3.5 Gy). This phenomenon was called the "Petcau effect". Similar effects were observed in studying biological effects under irradiation of plants in the region of the Chernobyl accident[197] and plants and animals after the Kystym accident in 1957.[198]

The "Petcau effect" was observed several times in different stochastic radiobiological events, such as the origin of cancer in plants, animals, and human beings.[199] Analysis of the health conditions of 200,000 men in Chernobyl (elimninating the Chernobyl accident consequences) for several groups of illnesses has shown a bimodal character of sick rate from the dose.[200] The maximum in the curve at small doses was observed, then this sick rate value was achieved at a dose five times larger. A large spread of sensitivity of biological objects to small dose (~20 times) was noticed. Synergistic effect under the action of radiation and other hazard factors of the environment was discovered, too. These effects are related to attenuation or unoperating of regeneration systems, but the mechanism of a small radiation dose effect on biological objects has not been explained so far.

9.4.5 Biofunctional and Immobilized Biofunctional Materials

Radiation techniques for producing biomaterials and materials used in implants allow products of higher purity level to be obtained (absence of catalysts, antidioxants, initiators, etc.) as compared to other methods. This reduces toxicological effects and, moreover, the production of biomaterials coincides with their sterilization.

The implants use silicon rubbers, polyethylenetherephthalate fibers, polyurethanes, etc.[201] To increase the biocoincidence of materials grafting of 2-hydroxyethylmetacrylate and ethylmetacrylate to polyethylene[202] and ethylpropylene rubber[203] was carried out. Adding hydrogels to the surface of hydrophobic polymer materials (grafting of hydrophyl monomers) sufficiently increases biocompatibility while maintaining the ability to fulfill relevant functions.

To improve compatibility with blood, grafting of acrylamide to polyurethane,[204] silicon rubbers,[205] and polyethylene[206] was used. Thrombo-resistivity is a property of polytetrafluoroethylene with grafted *N,N*-dimethylacrylamide.[207] Polyethylenetherephthalate fibers with grafted acrylamid are used to prepare artificial vessel limbs.[208,209] Radiation grafting of different monomers to a number of polymer substructures allows materials with hydrophyl surfaces possessing neutral, anion, or cation properties to be obtained.[210] The radiation method of immobilization in a surface is also used to create sorbents possessing ion-exchange complexing properties. Such sorbents are widely used in chromatography to analyze and separate biomolecules, in systems that rid the body of toxic substances (for example, heavy metals, etc.); in sensor devices; and in diagnostic tests.[211]

Radiation polymerization and polymer cross-linking are applied to preparing hydrogels in the form of solid gels or microparticles. Under radiation, cross-linking of polyvinyl alcohol, a hydrogel, occurs and can be used in artificial eyes.[212] To produce soft contact lenses, a hydrogel obtained by radiation polymerization of 2-hydroxyethylmetacrylate is used.[213]

Upon γ-radiation of maleic acid, acrylamide (a hydrogel) is formed in aqueous solution and absorbs bovine serum albumin.[214] Hydrophyl linear poly(acryloyl-L-proline methyl ester) gels can be used as controlled drug delivery systems.[215]

Immobilization (decrease in mobility, especially in solution) of bioactive materials (proteins, ferments, antibiotics, hormones, bacterial materials, etc.) is widely used to produce artificial organs, biosensors, "artificial kidney" systems, bioreactors, etc.[216] The main methods of immobilization are impregnation (impregnating a polymer matrix with active biomaterial), adsorption (physical and chemical), grafting to a substructure, and polymerization or cross-linking of bioactive materials. Under irradiation of solutions containing monomers and a bioactive substance, creation of a porous polymer matrix takes place which scavenges a bioactive component.[201,217,218] In this way, for example, medicinal preparations[209] and enzymes[219] have been immobilized.

Irradiation of such solutions at low temperatures allows a material with bioactive material immobilized on the surface of polymer pores to be obtained (due to localization of biocomponents in the ice-monomer interface surface).[220] To immobilize antibodies,[221] penicillin amylase,[222] and narciclasine,[223] methacrylate solutions frozen to –78°C were used.

Irradiation of a biocomponent-polymer mixture also leads to immobilization, since as a result of cross-linking a polymer matrix is formed which retains the biocomponents (for example, proteins and polypeptides).[224] Adsorption on different substructures was carried out to obtain materials with immobilized polysaccharide herapin[201] and yeast cells.[225,226] By use of solid polymerization of amino acids, initiated by γ-irradiation, a polymer poly(L-lencine) used in preparing temporary wound dressings is obtained.[227]

Grafting (formation of a covalence bond) of bioactive substances on a polymer substructure demands a presence on the substructure surface of reactivity groups which interact with the active center of the bioactive component (usually NH_2 group) to form a chemical bond. Such functional groups usually are OH or COOH which are grafted to a polymer if they are not present in its side chain. The structure of materials with grafted bioactive substances thus consists of a scavenger (support), anchor, (i.e., group responsible for bond stability of biocomponent with a scavenger), and legs (i.e., groups connecting the anchor with a biocomponent and biocomponents. To carry out reactions in "soft" conditions, different reactive groups are often added to the anchor. Examples of such immobilization are grafting of tyrpsin to polypropylene and polyvinylchloride;[228] L-asparaginase and monoclonal antibodies to polypropylene.[229,230] By means of immobilized L-asparaginase, the L-asparagine was removed from the blood of the patient when curing an acute lymphatic leukemia. Sepharose and Sephadex with grafted bovine serum albumin, and N-acryloxysuccinimide were used to separate D- and L-triptophans.[231] The addition of acrylamide to the NH_2 group of enzymes allows radiation grafting of the enzyme to cellulose with the acrylamide group acting as the anchor.[232] Examples of grafting different enzymes (papain, trypsin, etc.) to polyethylene are given in Reference 233.

REFERENCES

1. Woods, R.J., Pikaev, A.K., *Applied Radiation Chemistry: Radiation Processing,* Wiley-Interscience, New York, 1994, 535.
2. Samsonov, B.V., Tchyhanov, V.A., *Reactor Methods of Material Engineering,* Energoatomizdat, Moscow, 1991 (in Russian).
3. Kutuzov, V.L., Ovsyannikiv, M.Yu., Romanov, I.G., Pogrebuk, A.D., Remnev, G.E., Mechanical and frictional properties of tool steels exposed to HPIB irradiation, *Mech. Frictional Prop. Tool Steels,* November 8, 361–364, 1988.
4. Remnev, G.E., Shulov, V.A., *Practical Application of High-Power Ion Beams,* Paper 06-6, Beam 92, Washington, D.C., May 25-29, 1992.
5. Engelko, V.I., *Pulse Heat Treatment of Material Surface by High-Current Electron Beam,* Paper PH-50, Beam 92, Washington, D.C., May 25-29, 1992.
6. Proskurovsky, D.I., *Stress Waves Energy Dissipation and Structural Modification in Steels under the Action of a Low-Energy High-Current Electron Beam,* Paper PH-51c, Beam 92, Washington D.C., May 25-29, 1992.
7. Nesmelov, A.U., *Mechanical Properties of Metal and Alloys Exposed to High-Power Ion Beams,* Paper PH-54c, Beam 92, Washington, D.C., May 25-29, 1992.
8. Stiunett, R.W. et al., *The Use of Pulsed Intense Ion Beam for Thermal Surface Treatment,* 1993 IEEE International Conference on Plasma Science, Vancouver, B.C., Canada, June 7-9, 1993.
9. Kristiansen, M., *Pulsed Power Application,* 1993 IEEE International Pulsed Power Conference, New Mexico, June 21-23, 1993.
10. Hirvonen, J.K., Ed., *Ion Implanation,* Vol. 28, Academic Press, New York, 1980.
11. Dearnaey, G., Freeman, J.H., Nelson, R.S., Stephen, J., *Ion Implantation,* North-Holland, Amsterdam, 1973.
12. Smirnov, L.S., Ed., *Doping of Semiconductors by Means of Nuclear Reaction,* Nauka, Novosibirsk, 1981 (in Russian).
13. Kwast, H., Konrad, R., Elen, J.D., EXOTIC, an experimental programme on the development of ceramic thritium breeding material, *J. Nucl. Matter,* 246, 133–134, 1985.
14. Grigoriev, E.I., Trakhtenberg, L.I., Goldin, V.A. et al., Radiation thermal synthesis of oxide ceramics, in *Conf. Abstr. 11th Soviet Meeting on Theoretical and Applied Radiation Chemistry,* Obninsk, NITTEKhim, Moscow, Russia, 1990.
15. Gribkov, O.S., Voronin, A.P., The use of electron accelerators in the processes of high temperature solid phase synthesis, *Radiat. Phys. Chem.,* 46, 549, 1995.

16. Ibragimova, E.M., Gasanov, E.M., Polyak, O.Yu., Synthesis in γ-field and radiation treatment of high-Tc superconducting ceramics, in *Conf. Abstr. 9th Intl. Meeting on Radiation Processing,* Organizing Comnmittee, Istanbul, Turkey, 1994.
17. Koval, V.V., Kondratjev, V.A., Mischenko, A.I., Petrishghev, V.A., Rusakov, S.P., Radiation effect in catalysts production, in *Conf. Abstr. 9th Intl. Meeting on Radiation Processing,* Organizing Comnmittee, Istanbul, Turkey, 1994.
18. Bednoz, J.G., Muller, K.A., Possible high Tc superconductivity in barium-lantanum-copper-oxygen system, *Z. Phys.,* B64, 189, 1986.
19. Summers, G.P., Burke, E.A., Chrisey, D.B., Effect of particle-induced displacements of the critical temperature of $YBa_2Cu_3O_{7-d}$, *Appl. Phys. Lett.,* 55, 1469, 1989.
20. Kupfer, M., Wiech, U., Aptestend, I., Influence of fast neutron irradiation on inter- and intragrain properties of ceramic $YBa_2Cu_3O_{7-d}$, *IEEE Trans.,* 25, 2303, 1989.
21. Iwase, A., Watanabe, M., Iwata, T., Nihiru, T., Effect of 120 MeV ^{16}O ion irradiation at liquid nitrogen temperature on superconducting properties of Bi–Pb–Sr–Ca–Cu–O, *Jap. J. of Appl. Phys.,* 28, 1939, 1989.
22. Gous, F.K. Le., Ruault, M.O., Clerk, G. et. al., In situ studies of microstructural change in $YBa_2Cu_3O_{7-d}$ during O and He irradiation, *Phys. Magn.,* A60, 525, 1989.
23. Vichery, H., Rullier-Albenque, F., Koczukovski, M., Low temperature electron irradiation of $YBa_2Cu_3O_{7-d}$, *Physica C,* 162–164, 749, 1989.
24. Vorontsov, P.S., Grigoriev, E.I., Trakhtenberg, L.I. et al., The influence of γ-radiation on the properties of modified superconducting ceramics, *High Energy Chem.,* 28, 330, 1994.
25. Enokinara, A., Kohiki, S., Betsuni, K., X-ray irradiation effect on $ErYBa_2Cu_3O_{7-x}$ superconducting thin films, *Physica C,* 161, 431, 1989.
26. Soudr, E., Boatner, L.A., Chakoumakos, B.C., Sales, B.C., The effect of gamma irradiation on the superconducting properties of $GdYBa_2Cu_3O_{7-x}$, *Material Lett.,* 8, 171, 1989.
27. Kroto, H.W., Heath, J.R., O'Brien, S.C., Cure, R.F., Smalley, R.E., C_{60}; buckminster fullerene, *Nature,* 318, 162, 1985.
28. Kalish, R., Samoiloff, A., Hoffman, A., Uzan-Saguy, C., McCulloch, D., Prawer, S., Disintegration of C_{60} by heavy-ion irradiation, *Phys. Rev.,* B48, 18235(15), 1993.
29. Demirev, P., Papaleo, R.M., Eriksson, J., Brinkmalm, G., Ziegler, T., Hakansson, P., Sundqvist, B.U.P., Radiation-induced modification in C_{60} thin films, in *Proc. Symp. Recent Advances in the Chemistry and Physics of Fullerenes and Related Materials,* Kalish, K.M., Ruoff, R.S., Eds., The Electrochemical Society, Amsterdam, 1994, 397.
30. Rao, A.H., Zhou, P., Wang, K.A., Photoinduced polymerization of solid C_{60} films, *Science,* 259, 995, 1993.
31. Guldi, D.M., Hungerbuhler, H., Asmus, K.-D., Pulsed radiolysis and light-induced redox reactions of fullerenes on homo- and heterogeneous solution, in *Proc. Symp. on Recent Advances in the Chemistry and Physics of Fullerenes and Related Materials,* Kalish, K.M., Ruoff, R.S., Eds., The Electrochemical Society, Amsterdam, 1994, 854.
32. Unpublished results by Grigoriev, E.I., Trakhtenberg, L.I., Vorontsov, P.S., 1995.
33. Pace, M.D., Cristidis, T.C., Yin, J.J., Milliken, J., ESR of free radicals in C_{60}: effect of O_2, *J. Phys. Chem.,* 96, 6855, 1992.
34. Lubchenko, L.S., Kozhushner, M.A., Magnetic properties of polymers with system conjugated, *Russ. J. Phys. Chem.,* 62, 2308, 1988.
35. Arbogast, J.W., Darmanyan, A.P., Foote, C.S. et al., Photophysical properties of C_{60}, *J. Phys. Chem.,* 95, 11, 1991.
36. Bethune, D.S., Johnoson, R.D., Salem, J.R., deVries, M.S., Yannoni, C.S., Atoms in carbon cages: the structure and properties of endohedral fullerenes, *Nature,* 366, 123, 1993.

37. Haufler, R.E., Conceicao, J., Chibaute, L.P.F., Efficient production of C_{60} (buckminster fullerene), $C_{60}H_{36}$ and the solvated buckide ion, *J. Phys. Chem.,* 94, 8634, 1990.
38. Prassides, K., Ed., *Physics and Chemistry of the Fullerenes,* Kluwer Academic, Dordrecht, 1994.
39. Singh, A. Silverman, J., Eds., *Radiation Processing of Polymers,* Hanser, Munich, 1992.
40. Finkel, E.E., Leshenko, S.S., Braginsky, R.P., *Radiation Chemistry and Cable Processing,* Atomizdat, Moscow, 1968 (in Russian).
41. Silverman, J., Current status of radiation processing, *Radiat. Phys. Chem.,* 14, 1721, 1979.
42. Sakamoto, I., Mizusawa, K., Present state in progress of industrial electron processing systems in Japan, *Radiat. Phys. Chem.,* 22, 947, 1983.
43. Barlov, A., Brigge, J., and Maringer, M., Radiation processing of polyolefins and compounds, *Radiat. Phys. Chem.,* 9, 855, 1977.
44. Sagane, N., Harayama, H., Plastic foam-radiation cross-linked polyethylene foam, *Radiat. Phys. Chem.,* 18, 99, 1981.
45. Machi, S., Industrial application of radiation processing in Japan, *Radiat. Phys. Chem.,* 14, 155, 1979.
46. Charlesby, A., *Atomic Radiation and Polymers,* Pergamon Press, Oxford, 1960.
47. Ota, S., Current status of irradiated heat-shrinkable tubing in Japan, *J. Radiat. Chem. Phys.,* 18, 81, 1981.
48. Snyder, C.E., Keynote address, *J. Radiat. Chem. Phys.,* 22, 3, 1983.
49. Pekarskii, N.A., Osipov, B.V., Poddubbhyi, I.Ya. et al., Radiation-technological device RV-1200 for vulcanization of heatstable autoadhesion electroizolating tape and reinforced glass fiber rubber, *Atom. Energ.,* 36, 155, 1974.
50. Silverman, J., Basic concepts of radiation processing, *Radiat. Phys. Chem.,* 9, 1, 1977.
51. Tagawa, T., Tsuchida, E., Shinohara, M., Hagiwara, M., Kagiya, T., Accelerated radiation degradation of polymers by halogenated methane, *J. Polym. Sci. Polymer. Lett.,* 13, 287, 1975.
52. Ershov, B.G., Petruaev, E.P., Prospectives of use of radiation technology in the production of fodder, *High Energy Chem.,* 19, 3–489, 1985.
53. Leonhardt, J., Arnold, G., Baer, H., Radiation degradation of cellulose, *Radiat. Phys. Chem.,* 25, 899–904, 1985.
54. Kumamura, M., Kaexgy, I., Effect of electron beam current on radiation pretreatment of cellulosic wastes with electron beam accelerator, *Radiat. Chem. Phys.,* 23, 523–517, 1984.
55. Baer, H., Huebner, G., Leonhardt, J., *Biomass,* 2, 96, 1982.
56. Fleischer, R.L., Price, P.B., Walker, R.M., *Nuclear Tracks in Solids.* University of California Press, Berkeley, CA, 1975.
57. Flerov, G.N., Synthesis of superheavy elements and use of nuclear physics methods in adjacent fields, *Herald Russ. Acad. Sci.,* 4, 35–48, 1984.
58. Vater, P., Production and applications of nuclear track microfilters, *Nucl. Tracks Radiat. Meas.,* 15, 743–749, 1988.
59. Apel, P.Yu., Shirkova, V.V., Soboleva, T.I., Polypropylene nuclear membranes — new kind of membranes for decontamination and analysis of aggressive medium, *Vysokochistye Veshestva,* 2, 105, 1990.
60. Kaetsu, I., Yoshida, K., Cast polymerization of irradiation, *Radiat. Phys. Chem.,* 14, 737–745, 1979.
61. *Radiation Curing of Polymers,* Randell, D.R., Ed., The Royal Society of Chemistry, London, 1988.
62. Lauppi, U.V., Radiation curing — an overview, *Radiat. Phys. Chem.,* 35, 30, 1990.

63. *Proc. Radiation Curing,* Conference Rad. Tech. Asia Organizing Committee, 1991, Osaka, April 15-18, 1991.
64. Guimon, C., Radiation grafting of high density polyethylene and polypropylene, *Radiat. Phys. Chem.,* 14, 841, 1979.
65. Walsh, W.K., Oraby, W., Rucker, J., Radiation processes for textiles: curing of fabrics printed with pigmented foams and emulsion, *Radiat. Phys. Chem.,* 18, 253–265, 1981.
66. Stannett, V., Walsh, W.K., Bittencourt, E., Liepins, R., Surles, J.R., Chemical modification of fibers and fabrics with high-energy radiation, *Appl. Polym. Symp.,* 31, 201, 1977.
67. Walsh, W.K., Bittencourt, E., Radiation processes for textiles, *Intl. J. Radiat. Phys. Chem.,* 9, 289, 1977.
68. Hartley, F.A., Radiation grafting on polypropylene to prepare metal complex catalysts, *Br. Polym. J.,* 16, 199, 1984.
69. Rosenblyum, N.D., Radiation chemical synthesis of ion-exchange membranes enlargement scale, *Khim. Vysok. Energii,* 8, 339, 1974.
70. Kojuma, K., Kumafui, H., Ueno, K., Discoloration of plasticized PVC by irradiation, *Radiat. Phys. Chem.,* 18, 859, 1981.
71. Hegazy, E.A., El-Assy, N.B., Taher, N.H. et al., Radiation initiated grafting onto fluoropolymers for membrane preparation. II. *Radiat. Phys. Chem.,* 33, 539, 1989.
72. Bozzi, A., Chapiro, A., Synthesis of perm-selective membrane by grafting acrylic acid into air-irradiated teflon-FEP films, *Radiat. Phys. Chem.,* 32, 193, 1988.
73. Ivanov, A.E., Saburov, V.V., Zubov, V.P., Modification of mineral substrate by oligomers and polymers — way of synthesis of chromatography sorbents for biopolymers, *J. Sov. Chem. Soc.,* 34, 3, 1989.
74. Garnett, J.L., Grafting, *Radiat. Phys. Chem.,* 14, 79, 1979.
75. Stannett, V.T., Grafting, *Radiat. Phys. Chem.,* 18, 215, 1981.
76. Stannett, V.T., Radiation grafting — state-of-the-art, *Radiat. Phys. Chem.,* 35, 82, 1990.
77. Stannett, V.T., Silverman, J., Gartett, G.L., *Comprehensive Polymer Science,* Allen, G., Ed., Pergamon Press, New York, 1989, 317.
78. Borisov, E.N., Raichuk, F.Z., Miryeva, G.V., *Technology of Radiation-Chemical Producers of Cermete-Polymer and Wood-Polymer Materials,* Energoatomizdat, Moscow, 1982 (in Russian).
79. Witt, A.E., Henice, P.O., Griest, L.W., Acrylic woods in the United States, *Radiat. Phys. Chem.,* 18, 67, 1981.
80. Chan, K.Y., Yap, M.G.S., Chia, L.H.L., Neoh, K.G., Thermal characterization of rubberwood-polymer composites, *Radiat. Phys. Chem.,* 33, 197, 1989.
81. Meleshevich, A.P., *Methods of Radiation Chemistry in Production and Modification of Paper,* Energoatomizdat, Moscow, 1983 (in Russian).
82. Schaudy, R., Radiation-curable impregnates for the consolidation of wooden tins and art objects, *Radiat. Phys. Chem.,* 35, 71, 1990.
83. Dmitrenko, A.V., Ivanchev, S.S., *Reactions in Polymers Systems,* Khimia, Leningrad, 1987 (in Russian).
84. Tolstaya, S.N., Classification of surface-active modification by effective action in polymers, in *Problems of Polymers Composite Materials,* Naukova Dumka, Kiev, 1979 (in Russian).
85. Croker, R., Shneider, M., Hamman, K., Polymer reaction on surface powders, *Russ. Chem. Rev.,* 43, 2, 349, 1974.
86. Popov, V.À., Guzeev, V.V., Zvereva, Yu.À., Polymerization modification of fillers and its influences on properties of filled polyvinylchloride, *Polym. Sci.,* 1126, 1984.

87. Bruk, M.A., Pavlov, S.A., *Polimerization of Surface Solids,* Khimia, Moscow, 1990 (in Russian).
88. Bruk, M.A., *Radiation Vapour-Phase Encapsulation of Solids,* Energoatomizdat, Moscow, 1988 (in Russian).
89. P'yancov, G.N., in *Radiation Chemistry and Technology of Monomers and Polymers,* collection of scientific paperts, Naukova Dumka, Kiev, 1985 (in Russian).
90. Sykes, G.F., Bowles, D.E., Space radiation effects on the dimensional stability of a toughened epoxy graphite composite, *SAMPE Q.,* 17, 19, 1986.
91. Milhovich, S.M., Harakovich, C.T., Sikes, D.F., Space radiation effect on the thermo-mechanical behavior of graphite-epoxy composites, *J. Comp. Mater.,* 20, 519, 1986.
92. Harakovich, C.T., Fox, D.J., Sykes, G.T., *Trans. ASME:* Degradation of graphite-epoxy due electron radiation, *J. Eng. Mater. Technol.,* 110,2, 146–152, 1988.
93. Egusa, S.J., Kirk, M.A., Birtcher, R.C., Hagiwara, M., Kawanishi, S., Mechanical properties of organic composite materials irradiated with 2 MeV electrons, *J. Nucl. Mater.,* 119, 146, 1983.
94. Egusa, S.J., Mechanism of radiation-induced degradation in mechanical properties of polymer matrix composites, *Mater. Sci.,* 23, 2753, 1988.
95. Logunov, V.A., Role of collective effects in fluctuative dissociation anharmonic atomic chain, *Sov. Phys. Solid. Stat.,* 28(11), 3466–3472, 11986.
96. Egusa, S.J., Mechanism of radiation-induced degradation in mechanical properties of polymer matrix, *J. Mater. Sci.,* 23, 2753–2760, 1988.
97. Takeda, N., Tohdon, M., Takehashi, K., *SAMPE Quart.,* 20, 22, 27–32, 1989.
98. Seo, K.S., Forner, R.F., Gilbert, R.D., Effects of ionizing radiation on epoxy, graphite fiber, and epoxy/graphite fiber composites. Part I. Surface energy changes, *J. Polym. Sci.,* (Part B: *Polym. Phys.,*) 26, 245–255, 1988.
99. Vorontsov, P.S., Grigoriev, E.I., Trakhtenberg, L.I., Effect of radiation on the adhesion strength of composite materials, *Dokl. Phys. Chem.,* 309(6), 1386–1390, 1989.
100. Beklemishev S.À., Grigoriev, E.I., Mikheikin, I.D., Trakhtenberg, L.I., Influence of structure commensurability on the adhesional interaction, *Poverhnost,* 2(1), 19–24, 1989.
101. Beklemishev, S.À., Grigoriev, E.I., Trakhtenberg, L.I., Role of commensurability of elastic interacting surface at adhesion and deformation of the lattice, *Poverhnost,* 6(1), 10–14, 1991.
102. Kostikov, V.I., Carbon-carbon composite materials, *J. Sov. Chem. Soc.,* 34, 492, 1989.
103. Ahiezer, I.A., Davydov, L.N., *Introduction into Theoretical Physics, Metals and Alloys,* Naukova Dumka, Kiev, 1985 (in Russian).
104. Christensen, H., Effect of water radiolysis on corrosion in nuclear reactors, *Radiat. Phys. Chem.,* 18, 147, 1981.
105. Ishigure, K., Radiation chemistry related to nuclear power technology, *Radiat. Phys. Chem.,* 22, 119, 1983.
106. Burns, B.G., in *Radiation Research,* Vol. 2, Fielden, E.M. et al., Eds., Taylor & Francis, London, 1988, 193.
107. Uchida, S., Ibe, E., Katsura, R., Effect of irradiation on corrosion circumstances in boiling water, *Radiat. Phys. Chem.,* 22, 515, 1983.
108. Ishigure, K., Ikuse, H., Oshima, K. et al., The effect of radiation on the release of steel in high temperature water. II. Effect of flow rate and duration of corrosion experiment, *Radiat. Phys. Chem.,* 21, 281, 1983.
109. Campbell, F.J., Radiation damage in organic materials, *Radiat. Phys. Chem.,* 18, 109, 1981.
110. Gavrila, D., Paun, J., Influence of ionizing radiation on rheological, mechanical and electrical properties of various sorts of natural and synthetic rubber, *J. Radioanalyt. Nucl. Chem.,* 125, 243, 1988.

111. Ito, M., Effect of radiation on mechanical properties of polymers, *Proc. Intl. Symp. of Radiation, Degradation of Polymers and Radiation Resistant Materials,* Takasaka JAERI, Japan, 217–229, 1989.
112. Sasuga, T., Kawanishi, S., Seguchi, T., Kohno, I., Proton irradiation effects on several organic polymers, *Polymer,* 30, 2054, 1989.
113. Sasuga, T., Kawanishi, S., Nishii, M., Influence of types of ionizing radiation on radiation damage of polymers, 65–80, 1989.
114. Gillen, K.T., Clough, R.L., Techniques for monitoring heterogeneous oxidation of polymers, in *Handbook of Polymer Science and Technology,* Marcel-Dekker, New York, 1989, 2, 167.
115. Clough, R.L., Gillen, K.T., Heterogeneous oxidation of polymers by radiation, *Rad. Resis. Mat.,* 1989, 13–29.
116. Wilsky, H., The radiation induced degradation of polymers, *Radiat. Phys. Chem.,* 29, 1–14, 1987.
117. Tyutnev, A.P., Vannikov, A.V., Mingaleev, G.G., *Radiation Electrophysics of Organic Dielectrics,* Energoatomizdat, Moscow, 1989 (in Russian).
118. Gillen, K.T., Clough, R.L., Time temperature dose — rate superposition —a methodology for extrapolating accelerated radiation aging data to low-dose, *Polym. Degrad. Stab.,* 24, 137, 1989.
119. Burnay, S.G., Radiation effect on electrical strength of insulating materials, *Proc. Intl. Symp. Rad. Degr.,* Polym., Takasaki JAERI, Japan.
120. Tabata, Y., Iti, Y., Tagawa, S., Eds., *Handbook of Radiation Chemistry,* CRC Press, Boca Raton, FL, 1951.
121. Milinchuk, V.K., Tupicov, V.I., Eds., *Organic Radiation Chemistry Handbook,* Ellis Horwood, Chichester, U.K., 1989.
122. Egorov, G.F., *Radiation Chemistry of Extraction Systems,* Energoatomizdat, Moscow, 1986 (in Russian).
123. Horwitz, E.P., Dietz, M.L., Fisher, D.E., Correlation of the extraction of strontium nitrate by a crown-ether with the water-content of the organic phase, *Solv. Extract. Ion Exch.,* 8, 199, 1990.
124. Horwitz, E.P., Dietz, M.L., Fisher, D.E., Strex — a new process for the extraction and recovery of strontium from acidic nuclear waste streams, *Solv. Extract. Ion Exch.,* 9, 1, 1991.
125. Hooper, E.W., The Application of Inorganic Ion Exchanges to the Treatment of Medium Active Effluence, in *Solvent Extraction and Ion Exchange in the Nuclear Fuel Cycle,* Logsdail, D.H., Mills, A.L., Eds., Horwood Limited, Chichester, 1985, 145–155.
126. Bradbarg, D., Broadbent, Fitchett, G.A. et al., Ion exchange developments in management of CEGB full posds and associated wastes, in *Solvent Extraction and Ion Exchange in the Nuclear Fuel Cycle*, Logsdail, D.H., Mills, A.L., Eds., Ellis Horwood Limited, Chichester, 1985, 198–207.
127. Sato, K., Effect of seawater temperature on uranium recovery from seawater using amidoxime adsorbents, *Ind. Eng. Chem. Res.,* 33, 662–666, 1994.
128. Hiraoka, M., *Crown Compounds. Their Characteristics and Applications,* Kodansha, Tokyo, 1982.
129. Grigoriev, E.I., Nesterov, S.V., Trakhtenberg, L.I., Complexing ability of immobilized crown-ethers. I. Regularity of sorption ions, *Russ. J. Phys. Chem.,* 1995.
130. Grigoriev, E.I., Nesterov, S.V., Trakhtenberg, L.I., Complexing ability of immobilized crown-ethers. II. Effect of medium on complexing processes, *Russ. J. Phys. Chem.,* 1995.
131. Slovic, P., Flynn, J.H., Lauman, M.L., Perceived risk, trust, and the politics of nuclear waste, *Science,* 254, 1603–1607, 1991.

132. Report to the American Society by Waste Management, *Rev. Mod. Phys.,* 50, 1978.
133. Wicks, G.G., Nuclear Waste Glasses: Corrosion Behavior and Field Tests, in *Corrosion of Glass, Ceramics and Ceramic Superconductors, Principles, Testing, Characterization and Applications,* Clark, D.E., Ed., Zoitos, B.K., Eds., Noyes Publication, New Jersey, 1992.
134. *Scientific Basis for Nuclear Waste Management,* Lutze, W., Ewing, R.C., Eds., Vol. XII. Material Research Society, 1988, 127.
135. Makuuchi, K., Yoshi, F., Hyakutake, K., Feasibility study on utilization of vitrified waste as radiation source for processing, in *Conf. Abstr. 9th Intl. Meeting on Radiation Processing,* Organizing Committee, Istanbul, Turkey, 1994.
136. Myrova, L.O., Cheligenko, A.Z., *Support of Device Connection Stability under Ionizing and Electromagnetizing Irradiation,* Radio i Svyaz', Moscow, 1988 (in Russian).
137. Milintchuk, V.K., Klinshpont, E.R., Tupikov, V.I., *Base of Radiation Resistance of Organic Materials,* Energoatomizdat, Moscow, 1994 (in Russian).
138. Vavilov, V.S., *Effect of Radiation on Semiconductors,* Fizmatgiz, Moscow, 1963 (in Russian).
139. Dienes, G.J., Vineyard, G.H., *Radiation Effects in Solids,* Wiley-Interscience, New York, 1957.
140. Vavilov, V.S., Ukhin, N.A., *Radiation Processes in Semiconductors and Semiconductor Devices,* Atomizdat, Moscow, 1969 (in Russian).
141. Larin, F., *Radiation Effects in Semiconductor Devices,* McGraw-Hill, New York, 1968.
142. Brewer, G.R., Ed., *Electron Beam Technology in Microelectronics Fabrication,* Academic Press, London, 1980.
143. Baru, V.G., Volkenstein, F.F., *Irradiated Effect on Surface Properties of Semiconductors,* Nauka, Moscow, 1978 (in Russian).
144. Ladygin, E.A., Ed., *Effect of Penetrating Radiation on Electronic Devices,* Soviet Radio, Moscow, 1980 (in Russian).
145. Adams, L., Laboratory simulation of the space radiation environment for electronic components, *Radiat. Phys. Chem.,* 15, 525, 1980.
146. Carlson, R.O., Sun, Y.S., Assalit, H.B., Lifetime control in silicon power devices by electron or gamma irradiation, *IEEE Trans. T-ED,* 1103, 1977.
147. Williams, C.B., Vroom, D., A linear acceleration radiation processing facility, *Radiat. Phys. Chem.,* 14, 509, 1979.
148. Tarnega, K.S., Johnson, J.E., Bartko, J., US Patent 4.075.037 (21 Feb. 1978).
149. Fuochi, P.G., Hartelli, A., Combia, F. et al., Current status and prospects of electron irradiation of power semiconductor devices in Italy, *Radiat. Phys. Chem.,* 35, 493, 1990.
150. Sheng, W.V., Suh, Y.S.E., Tefft, E.G., German Patent 2837762 (15 May 1979).
151. Bartko, J., Chu, C.K., Schlegel, E.S., Tarnega, K.S., US Patent 4.3211.534 (19 Jan. 1982).
152. Sapienza, S., Fuochi, P.G., Switching time control on power high voltage bipolar transistors for high determination VDT by electron irradiation, *Radiat. Phys. Chem.,* 46, 1317, 1995.
153. Zhai, X.L., Chew, G.C., Qu, M.M., Radiation Processing of Power Semiconductor Devices in CIAE, in *Conf. Abstr. 9th Intl. Meeting on Radiation Processing,* Organizing Committee, Istanbul, Turkey, 1994.
154. Fuochi, P.G., Martelli, A., Bisio, G.H., Zitti, E.D., Motto, M.G., Passerini, B., Zambelli, M., Recombination centers and electrical characteristics in silicon power P-1-N diodes, *Radiat. Phys. Chem.,* 31, 809, 1988.
155. Prazil, Z., Schweiner, Z., Pesek, M., Radiation modification of physical properties of inorganic solids, *Radiat. Phys. Chem.,* 35, 509, 1990.

156. *Materials for Microlithography,* Advances in Chemistry Series, No. 266, American Chemical Society, Washington, D.C., 1985.
157. Morfeau, W.M., *Microlithography, Physics and Fabrication Technologies,* Plenum Press, New York, 1988.
158. Schnabel, M., Sotobayashi, H., Polymers in electron beam and x-ray lithography, *Prog. Polym. Sci.,* 9, 297–365, 1983.
159. Tabata, Y., Tageuva, S., Wasio, W., Study on sensitized degradation and crosslinking of polymers by means of picosecond pluse radiolysis, *Radiat. Phys. Chem.,* 25, 305, 1985.
160. Miller, L.J., Brault, R.G., Granger, D.D., Jensen, J.E., Vanast, C.I., Lewis, M.M., Poly(methacrylic anhydride) positive electron beam resist, *J. Vacuum Sci. Technol. B,* 7, 68, 1989.
161. Fischer, B.E., Spohr, R., Ion tracks and microstructure technology. Nuclear tracks and radiation measurements, *Intl. J. Rad. Appl. Instrum., Part D,* 15, 75, 1988.
162. Bowden, W.J., Lithography, *Radiat. Phys. Chem.,* 18, 357, 1981.
163. Fischer, B.E., Genswsurger, B., Spohr, R., Heavy ion lithography: a tool for the investigation and replication of microscopic objects, *Intl. J. Appl. Radiat. Isot.,* 31, 297, 1980.
164. Reichmanis, E., Thompson, L.F., Polymer materials for microlithography, *Chem. Rev.,* 89, 1273, 1989.
165. Personal correspondence with Bruk, M.A., 1995,
166. Nicelis, G., Prigoginl, Z., *Exploring Complexity, An Introduction,* Freeman, New York, 1989.
167. *Effect of Ionizing Radiation on DNA: Physical, Chemical and Biological Aspects,* Hütterman, Y., Köhnlein, W., Feule, R., Eds., Springer-Verlag, Berlin, 1978.
168. Mee, L.K., Radiation chemistry of biopolymers, in *Radiation Chemistry, Principles and Applications,* Farhataziz and Rodgers, M.A.J., Eds., VCH, New York, 1987, 477–499.
169. Rubin, A.B., *Biophysics. I. Theoretical Biophysics,* Vyshaya Shkola, Moscow, 1987 (in Russian).
170. Li, K.H., in *Recent Advances in Biophoton Research,* Popp, F.A., Ed., Word Science, Singapore, 1992, 157–195.
171. Lay, F.J.I., *Manual on Radiation Sterilization of Medical and Biological Materials,* International Atomic Energy Agency, Vienna, 1973, 37–64.
172. Tallentire, A., The spectrum of microbial radiation sensitivity, *Radiat. Phys. Chem.,* 15, 83, 1980.
173. Tallentire, A., Radiation on microbiology relevant to radiation processing, *Radiat. Phys. Chem.,* 4, 225, 1979.
174. Handlos, V., Sterilization by electron beam, *Radiat. Phys. Chem.,* 18, 175, 1981.
175. Takechisa, M., International standard (ISO) of radiation sterilization and issues in the sterilization dose selection, *Radiat. Phys. Chem.,* 46, 1349, 1995.
176. Tabei, M., Sekiguchi, M., Evaluation of sterilization dose for disposable hypodermic needles manufactured under ultra-clean conditions, *Radiat. Phys. Chem.,* 46, 629, 1995.
177. Whitby, J.L., Radiation resistance of *Acinetobacter* sp., *Radiat. Phys. Chem.,* 46, 639, 1995.
178. Landfield, H., The effects of radiation on various materials and the qualification test required for their use in medical devices, *Radiat. Phys. Chem.,* 15, 39, 1980.
179. Sklius, W.E., Sterilizing radiation effects on selective polymers, *Radiat. Phys. Chem.,* 15, 47, 1980.

180. Gopal, N.G.S., Patel, K.M., Sharma, G., Bhalla, H.L., Wills, P.A., Hilmy, N., Guide for radiation sterilization of pharmaceuticals and decontamination of raw materials, *Radiat. Phys. Chem.,* 32, 619, 1988.
181. Talrose, V.L., Trofimov, V.I., Problems of radiation sterilization, *High Energy Chem.,* 17, 233, 1983.
182. Talrose, V.L., Trofimov, V.I., Crioradiation sterilization — contemporary state and outlook, *Radiat. Phys. Chem.,* 46, 633, 1995.
183. Edwards, H.E., Philipse, G.O., Radiation effects of human tissues and their use in tissue banking, *Radiat. Phys. Chem.,* 22, 889, 1983.
184. Cleghorn, D.A., Nablo, S.V., Electron sterilization valuation techniques using the controlled depth of sterilization process, *Radiat. Phys. Chem.,* 35, 382, 1990.
185. Cleghorn, D.A., Nablo, S.V., Low energy electron surface dose measurements in pouched syringes, in *Conf. Abstr. 9th Intl. Meeting on Radiation Processing,* Organizing Committee, Istanbul, Turkey, 1994.
186. Dorpema, J.W., Risk assessment of medical devices: evaluation of microbiological and toxicological safety, in *Conf. Abstr. 9th Intl. Meeting on Radiation Processing,* Organizing Committee, Istanbul, Turkey, 1994.
187. *Irradiation of Foods,* May, J.H., Ed., Elsevier, London, 1989.
188. Johnston, D.E., Stevenson, M.H., Eds., *Food Irradiation and Chemistry,* Royal Society of Chemistry, Cambridge, 1950.
189. Kuzin, A.M., Kaushansky, D.A., *Applied Radiobiology,* Energoatonizdat, Moscow, 1981 (in Russian).
190. *Isotopes in Everyday Life,* International Atomic Energy Agency, Vienna, 1990, 18.
191. *Training Manual on Food Irradiation Technology and Techniques,* International Atomic Energy Agency, Vienna, 1982.
192. Loaharunnu, P.L., *International Atomic Energy Agency Yearbook, 1989,* International Atomic Energy Agency, Vienna, 1989, 135.
193. McKinley, R.C., Report on the activities of the international consultative group on food irradiation, *Radiat. Phys. Chem.,* 46, 745, 1995.
194. Loaharanu, P.L., Commercialization of food irradiation, *IAEA Bull.,* 32(2), 44, 1990.
195. ICRP, *Radiation Protection,* Publ. 60, *Ann. ICRP,* 21, 1–3, 1990.
196. Petkou, A., Radiation effect with a model lipid membrane, *Can. J. Chem.,* 49, 1187, 1971.
197. Abramov, V.I., Fedorenko, O.M., Shevehenko, V.A., Genetic consequences of radioactive contamination for population of Arabidopsis, *Sci. Total Environ.,* 112, 19–29, 1992.
198. Shevehenko, V.A., Pechkurenkov, V.L., Abramov, V.I., Study of genetic effects induced in population by radioactive nuclear decay products U-239. Communication II. Prediction of genetic effectivity radiation at low rate dose, *Russ. J. Genet.,* 14, 622, 1978.
199. Graeub, R., *The Petkou Effect of Nuclear Radiation, People and Trees,* Four Walls Eight Windows, New York, 1992, 232.
200. *Chernobyl Catastrophe: Reasons and After-Effects (Expert Conclusion),* Part 2. *Medico-Biological and Genetic After-Effects by Chernobyl Catastrophe,* Burlacova, E.B., Ed., Nauka i Technika, Minsk, 1993 (in Russian).
201. Hoffman, A.S., A review of the use of radiation plus chemical and biochemical processing treatments to prepare mobile biomaterials, *Radiat. Phys. Chem.,* 18, 323–342, 1981.
202. Hoffman, A.S., Application of radiation-grafted hydrogels as blood-contacting biomaterials, *Radiat. Phys. Chem.,* 22, 267–283, 1983.
203. Anelli, P., Bassaro, S., Carenza, M., Palma, G., Radiation grafting of hydrophilic monomers onto ethylene-propylene rubber, *Radiat. Phys. Chem.,* 46, 1031, 1995.

204. Jancen, B., Ludwika, A., Storz, L.W., Radiation included modification of polyether-urethane films and tubes: platelet adhesion and in vivo experiments, *Radiat. Phys. Chem.,* 25, 529, 1985.
205. Fugioka, S., Fugikawa, K., Ueno, N., Okamoto, A., Electron beam curing coating and process for high-grade precoated steel sheet, *Radiat. Phys. Chem.,* 19, 865, 1981.
206. Ikada, Y., Suzuki, M., Taniguchi, M., Interaction of blood with radiation-grafted materials, *Radiat. Phys. Chem.,* 18, 207, 1981.
207. Otsuhata, K., Pazzak, M.T., Castanares, R.L., Tabata, Y., Ohashi, F., Takeuchi, A., Effect of surface texture of grafted films on antithrombogenicity, *Radiat. Phys. Chem.,* 25, 537, 1985.
208. Ruciuska, A., Rosiak, J., Pekula, W., Radiation modification of vascular prostheses, *Radiat. Phys. Chem.,* 24, 495, 1984.
209. Pekula, W., Rosiak, J., Ruciuska-Rybus, A., Radiation crosslinked hydrogels as sustained release drug delivery systems, *Radiat. Phys. Chem.,* 27, 275, 1986.
210. Lora, S., Garenza, M., Palma, G. et al., Radiation-induced graft co-polymerization of hydrophilic monomers onto polyphosphazenes, *Radiat. Phys. Chem.,* 35, 117, 1990.
211. Janata, J., Josowicz, M., DeVaney, D.M., Chemical sensors, *Anal. Chem.,* 66, 207R, 1994.
212. Hara, Y., Kumiya, S., Nishioka, K. et al., Electron beam irradiation conditions and foamed seat physical properties in polypropylene-polyethylene blends, *Acta Soc. Ophthalmol. Jap.,* 83, 1478, 1979.
213. Ming, L.Yu., Qi, Y.Y., Jil, C. et al., Study of the kinetics and influence factor of radiation synthesis of hydrogel material, *Radiat. Phys. Chem.,* 35, 122, 1990.
214. Saraydin, D., Karadag, E., Oztop, N., Given, O., Preparation of acrylamide-mallic acid hydrogels and their use in the absorbtion of bovine serum albumin, *Radiat. Phys. Chem.,* 46, 1053, 1995.
215. Yoshida, M., Safrang, A., Omichi, H. et al., Intelligent biomedical gels based on pendant L-proline alkyl esters, in *Conf. Abstr. 9th Intl. Meeting on Radiation Processing,* Organizing Committee, Istanbul, Turkey, 1994.
216. Hoffman, A.S., *Radiation Technology for Immobilization of Biactive Materials,* International Atomic Energy Agency, Vienna, 1988, 25.
217. Kawashima, K., Umeda, K., Immobilization of enzymes by radiopolymerization of acrylamide, *Biotechnol. Bioeng.,* 16, 609, 1974.
218. Kawashima, K., Umeda, K., Immobilization of enzymes by radiopolymerization of acrylamide, *Biotechnol. Bioeng.,* 17, 599, 1975.
219. Kaetsu, I., Kumamura, M., Immobilization of enzymes and antibodies by low energy electron beam polymerization, *Radiat. Phys. Chem.,* 30, 263, 1987.
220. Kaetsu, I., Immobilization biofunctional substance, *Radiat. Phys. Chem.,* 18, 343, 1981.
221. Kaetsu, I., Kumamura, M., Fugimura, T., Studies on the immobilization of biofunctional components by radiation polymerization and applications, *Radiat. Phys. Chem.,* 27, 245, 1986.
222. Garenza, M., Cora, S., Palma, G., Bocci, E., Largajolli, R., Veronese, F.M., Influence of matrix porosity on the immobilization of penicillin acylase by radiation, *Radiat. Phys. Chem.,* 31, 657, 1988.
223. Veronese, F.M., Ceriotti, G., Keller, G., Controlled release of narciclasine from poly(2-hydroethylmethacrylate) poly (HEMA) matrixes polymerized by a chemical initiator and gamma irradiation, *Radiat. Phys. Chem.,* 35, 88, 1990.
224. Kaetsu, I., Recent progress on the immobilization of biocomponents by radiation polymerization and application to biomedical uses, *Radiat. Phys. Chem.,* 25, 517, 1985.

225. Kaetsu, I., Kumamura, M., Fugimura, T., Immobilization of microbial cell and yeast cell and application to biomass conversion using radiation techniques, *Radiat. Phys. Chem.,* 29, 191, 1987.
226. Fugimura, T., Garenza, M., Kaetsu, I., Activity dependence of immobilized cells on the various natures of polymer carriers for immobilization produced with radiation polymerization at low temperature, *Radiat. Phys. Chem.,* 31, 653, 1988.
227. Leh, C.-H., A study on the solid-state polymerization of poly(L-leucine) initiated by γ-ray, *Radiat. Phys. Chem.,* 46, 851, 1995.
228. Garnett, J.L., Jankewicz, S.V., Levot, R., Sangster, D.F., Insolubilization of biologically active materials with novel radiation graft copolymers, *Radiat. Phys. Chem.,* 25, 509, 1985.
229. Gombotz, W., Hoffman, A.S., Schmez, G., Venoya, M.O., Immobilization of enzymes in blood plasma exchanger via radiation grafting, *Radiat. Phys. Chem.,* 25, 549, 1985.
230. Hoffman, A.S., Gombotz, W., Uenoyama, S., Dong, L.C., Schmer, G., Immobilization of enzymes and antibodies to radiation grafted polymers. Therapeutic and diagnostic applications, *Radiat. Phys. Chem.,* 27, 265, 1986.
231. Jun, Li. Min, Yi, Hogfei, Ha., Application of sepharose and sephadex modified by means of radiation grafting in separation of biomolecules, *Radiat. Phys. Chem.,* 46, 847, 1995.
232. Dong, L.C., Hoffman, A.S., A new method for immobilization of biomolecules using pre-irradiation grafting at low temperature, *Radiat. Phys. Chem.,* 28, 177, 1986.
233. Silva, M.A., Gil, M.H., Guiomar, J., Lapa, E., Machado, E., Moreira, M., Guthrie, J.T., Kotov, S., The immobilization of enzymes onto poly(ethylene)-g.co-metacrylic acid, [poly(ethylene)-g.co-hydroxyethyl metacrylate]-g.co-metacrylic acid and [poly(ethylene)-g.co-hydroxyethyl metacrylate]-g.co-hydroxyethyl metacrylate, *Radiat. Phys. Chem.,* 36, 589, 1990.

INDEX

C

F

G

H

I

K

L

M

N

O

P

R

S